Springer

*New York
Berlin
Heidelberg
Barcelona
Budapest
Hong Kong
London
Milan
Paris
Santa Clara
Singapore
Tokyo*

Universitext

Editors (North America): S. Axler, F.W. Gehring, and P.R. Halmos

Aksoy/Khamsi: Nonstandard Methods in Fixed Point Theory
Aupetit: A Primer on Spectral Theory
Booss/Bleecker: Topology and Analysis
Borkar: Probability Theory: An Advanced Course
Carleson/Gamelin: Complex Dynamics
Cecil: Lie Sphere Geometry: With Applications to Submanifolds
Chae: Lebesgue Integration (2nd ed.)
Charlap: Bieberbach Groups and Flat Manifolds
Chern: Complex Manifolds Without Potential Theory
Cohn: A Classical Invitation to Algebraic Numbers and Class Fields
Curtis: Abstract Linear Algebra
Curtis: Matrix Groups
DiBenedetto: Degenerate Parabolic Equations
Dimca: Singularities and Topology of Hypersurfaces
Edwards: A Formal Background to Mathematics I a/b
Edwards: A Formal Background to Mathematics II a/b
Foulds: Graph Theory Applications
Fuhrmann: A Polynomial Approach to Linear Algebra
Gardiner: A First Course in Group Theory
Gårding/Tambour: Algebra for Computer Science
Goldblatt: Orthogonality and Spacetime Geometry
Hahn: Quadratic Algebras, Clifford Algebras, and Arithmetic Witt Groups
Holmgren: A First Course in Discrete Dynamical Systems
Howe/Tan: Non-Abelian Harmonic Analysis: Applications of $SL(2, R)$
Howes: Modern Analysis and Topology
Humi/Miller: Second Course in Ordinary Differential Equations
Hurwitz/Kritikos: Lectures on Number Theory
Jennings: Modern Geometry with Applications
Jones/Morris/Pearson: Abstract Algebra and Famous Impossibilities
Kannan/Krueger: Advanced Analysis
Kelly/Matthews: The Non-Euclidean Hyperbolic Plane
Kostrikin: Introduction to Algebra
Luecking/Rubel: Complex Analysis: A Functional Analysis Approach
MacLane/Moerdijk: Sheaves in Geometry and Logic
Marcus: Number Fields
McCarthy: Introduction to Arithmetical Functions
Meyer: Essential Mathematics for Applied Fields
Mines/Richman/Ruitenburg: A Course in Constructive Algebra
Moise: Introductory Problems Course in Analysis and Topology
Morris: Introduction to Game Theory
Porter/Woods: Extensions and Absolutes of Hausdorff Spaces
Ramsay/Richtmyer: Introduction to Hyperbolic Geometry
Reisel: Elementary Theory of Metric Spaces
Rickart: Natural Function Algebras
Rotman: Galois Theory

(continued after index)

R. Kannan   Carole King Krueger

Advanced Analysis

on the Real Line

Springer

R. Kannan
Carole King Krueger
Department of Mathematics
University of Texas at Arlington
Arlington, TX 76019-0408

Mathematics Subject Classification (1991): 26-01, 28Axx

Library of Congress Cataloging-in-Publication Data
Kannan, Rangachary.
    Advanced analysis on the real line / R. Kannan, Carole King Krueger.
       p.   cm. — (Universitext)
    Includes bibliographical references and index.

    1. Mathematical analysis.   2. Functions of real variables.
I. Krueger, Carole King.   II. Title.
QA300.K35   1996
515′.8—dc20   95-49238

Printed on acid-free paper.

Production managed by Natalie Johnson; manufacturing supervised by Jacqui Ashri.
Typeset by Asco Trade Typesetting Ltd., Hong Kong.
Printed and bound by R.R. Donnelley and Sons, Harrisonburg, VA.

9 8 7 6 5 4 3 2 1
ISBN-13: 978-0-387-94642-9   e-ISBN-13: 978-1-4613-8474-8
DOI: 10.1007/978-1-4613-8474-8

Preface

Since the publication of Saks' classic treatise, *Theory of the Integral*, most of the books published on analysis of functions of a single real variable have been at the introductory level. There have not been many attempts, to our knowledge, to introduce students to some of the topics of analysis in $\mathbf{R}^1$ that are fundamental to analysis in $\mathbf{R}^n$. Some of these important topics include density, approximate continuity, approximate differentiability and Hausdorff measure. As we now know, functions of bounded variation in $\mathbf{R}^n$ play one of the most important roles in current developments in calculus of variations and partial differential equations. Sard's theorem is essential to differential topology. Another topic which plays an important role in $\mathbf{R}^n$ is differentiability; the book by Saks essentially ends with a brief introduction to this topic.

Our original motivation was to write a book which dealt with these topics and illustrated their role in $\mathbf{R}^1$ and $\mathbf{R}^n$. However, we quickly discovered that the topics we needed to cover in $\mathbf{R}^1$ were sufficient to warrant a volume of their own. In the process of writing about these topics, we discovered a number of exciting and powerful results in $\mathbf{R}^1$ that are not found in most books. The results on Cantor sets and singular functions, Dini derivatives, monotone functions, and the Denjoy theorem illustrate the beauty of analysis in $\mathbf{R}^1$ in its own right. We have included a number of interesting results, largely from original sources, that have received little, if any, attention in textbooks. Some of these nonstandard topics are:

- Tolstov's theorems which generalize the familiar theorem that $f' > 0 \rightarrow f$ is increasing to give conditions on the approximate derivatives to insure the same conclusion

- Sierpenski's theorem, which extends the property of approximate continuity to arbitrary functions
- Ruziewicz's example of an infinite number of functions, which have equal derivatives (not finite everywhere) but the difference of any two is not a constant
- the Darboux property, also called the intermediate value property
- the Denjoy–Saks–Young theorem characterizing the Dini derivatives and the approximate derivatives of any function.

This book assumes the reader has a knowledge of introductory real analysis. Some introductory measure theory would also be helpful. A brief summary of the results from measure theory which are needed for this book is given in Chapter 0. The teacher could provide additional instruction for those students whose background is lacking. We have attempted to provide exercises of varying degrees of difficulty to help the reader understand the contents of each chapter and, in some cases, to extend the results of the text.

We realize that we may have omitted topics which others consider important. Our goal has not been to produce an exhaustive treatise on real analysis but rather to introduce topics which are important to analysis in $\mathbf{R}^n$ in their simplest setting in $\mathbf{R}^1$.

A sequel to this book is currently in progress and will present advanced topics in analysis in $\mathbf{R}^n$, such as:

- Sard's theorem
- Besicovitch covering theorem
- bounded variation in $\mathbf{R}^n$
- density in $\mathbf{R}^n$
- Hausdorff measure.

We would like to acknowledge the profound influence of Saks' classic work, whose spirit may be seen pervading our book almost everywhere. Additional motivation for this book was provided by Hobson's *The Theory of Functions of a Real Variable and the Theory of Fourier Series*. While we mention these two classics in particular, we are also indebted to a multitude of other books and journal articles which are listed in our bibliography and from which we have borrowed freely. We apologize for the exclusion of any reference or the failure to correctly attribute any result. We would like to acknowledge the help of numerous colleagues who read various parts of the manuscript. The Interlibrary Loan Office of the University of Texas at Arlington was invaluable in obtaining many of our references. Finally, we would like to thank our editor, Dr. Ina Lindemann, and the editorial staff at Springer-Verlag for their patience and guidance.

Contents

CHAPTER 0

Preliminaries

This chapter contains a summary of some of the important results in analysis which are used in this book. For a more complete treatment of these topics, the reader may consult [N1], [R4], or [T2].

§0.1. Lebesgue Measure

The classical theory of Lebesgue measure was presented by Henri Lebesgue in the early years of the twentieth century; his motivation was to develop a theory of integration more general than that of Riemann. In Riemann integration, we consider functions defined on intervals; the Lebesgue integral allows us to consider functions defined on arbitrary sets. The notion of measure generalizes the concept of the length of an interval (or for the case of sets in $\mathbf{R}^n$, the area of a rectangle, the volume of a parallelepiped, etc.).

Definition 0.1.1. A *set function* is a function which associates an extended real number to each set in some collection of sets. The length of an interval is an example of a set function defined on all intervals of the real line. If $I = [a, b]$, then $|I|$ shall be used to denote the length of I; thus, $|I| = b - a$.

Definition 0.1.2. If $E \subset \mathbf{R}^1$ and if $\{I_n\}$ is a countable collection of open intervals such that

$$E \subset \bigcup_n I_n,$$

then $\{I_n\}$ is called a (*Lebesgue*) *covering* of E. Note that there are many such coverings for a given set.

1

Definition 0.1.3. For a set $E \subset \mathbf{R}^1$ and a covering $\{I_n\}$ of E, consider the sum of the lengths of the intervals of the covering,

$$\sum_n |I_n|.$$

The *outer (Lebesgue) measure* of E, denoted $m^*(E)$, is the greatest lower bound of all such sums:

$$m^*(E) = \inf_{E \subset \cup I_n} \sum_n |I_n|.$$

It follows immediately from the definition of m^* that $m^*(E) \geq 0$; $m^*(\varnothing) = 0$; and if $A \subset B$, then $m^*(A) \leq m^*(B)$. We list the main properties of the outer measure m^*; the proofs can be found in the references cited at the beginning of the chapter.

0.1.4. The outer measure of an interval is its length.

0.1.5. Let I be any open interval and $\bar{I}$ its closure. Then $m^*(\bar{I}) = |I|$.

0.1.6. Let $\{E_n\}$ be any countable collection of subsets of $\mathbf{R}^1$. Then

$$m^*\left(\bigcup_n E_n\right) \leq \sum_n m^*(E_n).$$

0.1.7. If E is countable, then $m^*(E) = 0$. Sets of measure zero are also called *null* sets.

Definition 0.1.8. A set E is said to be *(Lebesgue) measurable* if for each set $A \subset \mathbf{R}^1$, we have

$$m^*(A) = m^*(A \cap E) + m^*(A \cap E^c),$$

where E^c is the complement of the set E. Since $(A \cap E) \cup (A \cap E^c) = A$, it follows from Property 0.1.6 that

$$m^*(A) \leq m^*(A \cap E) + m^*(A \cap E^c).$$

Hence, E is measurable if and only if for each set $A \subset \mathbf{R}^1$, we have

$$m^*(A) \geq m^*(A \cap E) + m^*(A \cap E^c).$$

Because the definition of measurability is symmetric in E, we have E^c is measurable whenever E is measurable. Obviously $\varnothing$ and the set $\mathbf{R}^1$ are measurable.

Definition 0.1.9. If E is a measurable set, the *(Lebesgue) measure* mE is defined as the outer measure of the set E. Thus, m is the set function obtained by restricting the set function m^* to the class of measurable subsets of $\mathbf{R}^1$.

Properties of Lebesgue measure:

0.1.10. If $m^*(E) = 0$, then E is measurable. Thus, if $F \subset E$ and $m^*(E) = 0$, then F is measurable.

0.1.11. If E_1 and E_2 are measurable, so are $E_1 \cup E_2$, $E_1 \cap E_2$, and $E_1 - E_2$.

0.1.12. The collection of measurable sets is a σ-algebra; that is, the complement of a measurable set is measurable and the union and intersection of a countable collection of measurable sets are measurable.

0.1.13. Every Borel set[1] is measurable. In particular, each open set and each closed set is measurable.

0.1.14. Let $\{E_n\}$ be a sequence of measurable sets. Then

$$m\left(\bigcup_n E_n\right) \le \sum_n mE_n.$$

If the sets $\{E_n\}$ are pairwise disjoint, then

$$m\left(\bigcup_n E_n\right) = \sum_n mE_n.$$

0.1.15. Let $\{E_n\}$ be an infinite decreasing sequence of measurable sets, i.e., $E_{n+1} \subset E_n$ for each n. Let mE_1 be finite. Then

$$m\left(\bigcap_{n=1}^{\infty} E_n\right) = \lim_{n \to \infty} mE_n.$$

0.1.16. Let $E \subset \mathbf{R}^1$. Then the following statements are equivalent:

(a) E is measurable.
(b) Given $\varepsilon > 0$, there is an open set $O \supset E$ with $m^*(O - E) < \varepsilon$.
(c) Given $\varepsilon > 0$, there is a closed set $F \subset E$ with $m^*(E - F) < \varepsilon$.
(d) There is a G_δ set G with $E \subset G$ and $m^*(G - E) = 0$.
(e) There is a F_σ set F with $F \subset E$ and $m^*(E - F) = 0$.

If $m^*(E)$ is finite, the above statements are equivalent to:

(f) Given $\varepsilon > 0$, there is a finite union U of open intervals such that $m^*(U \Delta E) < \varepsilon$. $U \Delta E$ denotes the *symmetric difference* of the two sets U and E; i.e.,

$$U \Delta E = (U - E) \cup (E - U).$$

---

[1] The collection B of Borel sets is the smallest σ-algebra which contains all of the open sets. A set which is a countable union of closed sets is called an F_σ. A set which is the intersection of a countable collection of open sets is called a G_δ.

0.1.17. For each set E, we have

$$m^*(E) = \inf\{mU: U \text{ open}, E \subset U\}$$

and

$$m^*(E) = \inf\{mF: E \subset F, F \text{ measurable}\}.$$

Outer measures which satisfy property (0.1.17) are called *regular*.

0.1.18. For any measurable set E,

$$mE = \sup\{mA: A \subset E, A \text{ compact}\}.$$

Theorem 0.1.19. *Let f be an extended real-valued[2] function whose domain is measurable. Then the following statements are equivalent:*

(a) *For each real number α, the set $\{x: f(x) > \alpha\}$ is measurable.*
(b) *For each real number α, the set $\{x: f(x) \geq \alpha\}$ is measurable.*
(c) *For each real number α, the set $\{x: f(x) < \alpha\}$ is measurable.*
(d) *For each real number α, the set $\{x: f(x) \leq \alpha\}$ is measurable.*

These statements imply

(e) *For each extended real number α, the set $\{x: f(x) = \alpha\}$ is measurable.*

Definition 0.1.20. An extended real-valued function f is said to be (*Lebesgue*) *measurable* if its domain is measurable and if it satisfies one of the first four statments of Theorem 0.1.19.

Theorem 0.1.21. *An extended real-valued function with measurable domain is measurable if and only if $f^{-1}(G)$ is measurable whenever G is an open set in $\mathbf{R}^*$ (with the usual topology).*

Remark. A function is *Borel measurable* if for each $\alpha \in \mathbf{R}^1$, the set $\{x: f(x) > \alpha\}$ is a Borel set. Theorem 0.1.19 holds if "measurable" is replaced by "Borel set."

Corollary 0.1.22. *A continuous function with measurable domain is measurable. A step function is measurable. If f is a measurable function and E is a measurable subset of the domain of f, then $f|_E$ (the function obtained by restricting f to E) is also measurable.*

Definition 0.1.23. If A is any set, the *characteristic function* χ_A of the set A is defined

$$\chi_A(x) = \begin{cases} 1, & x \in A \\ 0, & x \notin A. \end{cases}$$

---

[2] $\mathbf{R}^*$ denotes the extended real numbers, i.e., $\mathbf{R}^1 \cup \{\pm\infty\}$.

Corollary 0.1.24. *The characteristic function χ_A of a set A is measurable if and only if A is measurable.*

Theorem 0.1.25. *Let $c \in \mathbf{R}^1$ and f and g be two measurable real-valued functions defined on the same domain. Then the functions $f + c$, cf, $f + g$, $g - f$, and fg are also measurable.*

For extended real-valued functions f and g, $f + g$ is not defined at points of the form $\infty - \infty$. If we take the same value for $f + g$ at points where it is undefined, then $f + g$ will be measurable. The remainder of Theorem 0.1.25 can also be applied to extended real-valued functions.

Definition 0.1.26. Given an extended real-valued function f, we define the *positive part f^+ of f* by

$$f^+(x) = \begin{cases} f(x) & \text{if } f(x) \geq 0 \\ 0 & \text{if } f(x) < 0. \end{cases}$$

We define the *negative part f^- of f* by

$$f^-(x) = \begin{cases} -f(x) & \text{if } f(x) \leq 0 \\ 0 & \text{if } f(x) > 0. \end{cases}$$

Observe that

$$f = f^+ - f^-$$

and

$$|f| = f^+ + f^-.$$

Theorem 0.1.27. *If f is an extended real-valued measurable function, so are f^+ and f^-. If f^+ and f^- are measurable, so are f and $|f|$.*

Theorem 0.1.28. *Let $\{f_n\}$ be a sequence of measurable functions with the same domain. Then the functions $\sup\{f_1, \ldots, f_n\}$, $\inf\{f_1, \ldots, f_n\}$, $\sup_n f_n$, $\inf_n f_n$, $\overline{\lim} f_n$, and $\underline{\lim} f_n$ are measurable.*

Definition 0.1.29. If some property holds for all points in a set E except for a subset $E_0 \subset E$ such that $mE_0 = 0$, then that property is said to hold *almost everywhere* (abbreviated a.e.) on E. Thus, to say $f = g$ a.e. means the set of points $\{x : f(x) \neq g(x)\}$ has measure zero (of course, f and g have the same domain). If $f = g$ a.e., then f and g are called *equivalent functions*.

The relation $f = g$ a.e. is a true equivalence relation within the class of functions defined on given domain which is a subset of $\mathbf{R}^*$; that is, it is reflexive, symmetric, and transitive.

Theorem 0.1.30. *If f is a measurable function and $f = g$ a.e., then g is measurable.*

Theorem 0.1.31. *Let E be a measurable set of finite measure and $\{f_n\}$ a sequence of measurable functions which are defined and finite a.e. on E. Suppose f is a real-valued function such that $f_n(x) \to f(x)$ a.e. on E. Then*

$$\lim_{n \to \infty} m\{x: |f_n(x) - f(x)| \geq \delta\} = 0$$

for all $\delta > 0$.

Theorem 0.1.32 (Egoroff's Theorem). *If $\{f_n\}$ is a sequence of measurable functions that converge to a real-valued function f a.e. on a measurable set E of finite measure, then given $\eta > 0$, there is a subset $A \subset E$ with $mA < \eta$ such that f_n converges uniformly to f on $E - A$.*

A frequent tool used in analysis is that of approximation; for example, we approximate irrational numbers by rational numbers and continuous functions by polynomials, such approximation serving to make the task at hand somehow simpler. The next results deal with the approximation of measurable functions.

Theorem 0.1.33. *Let f be a measurable function defined on an interval $[a, b]$. Assume f takes the values $\pm\infty$ only on a set of measure zero. Then, given $\varepsilon > 0$, we can find a step function g and a continuous function h such that*

$$|f - g| < \varepsilon \quad and \quad |f - h| < \varepsilon$$

except on a set of measure less than ε; i.e., $m\{x: |f(x) - g(x)| \geq \varepsilon\} < \varepsilon$ and $m\{x: |f(x) - h(x)| \geq \varepsilon\} < \varepsilon$. If in addition $k \leq f \leq K$, then we may choose the functions g and h so that they are similarly bounded.

Theorem 0.1.34 (Lusin's Theorem). *Let $f: [a, b] \to \mathbf{R}^*$ be measurable and finite a.e. Then, for $\delta > 0$, there is a continuous function g defined on $[a, b]$ such that $m\{f(x) \neq g(x)\} < \delta$. The result is also true if $[a, b] = R^1$.*

Definition 0.1.35. A real-valued function ϕ is called *simple* if it is measurable and its range is a finite set of real numbers. If ϕ is a simple function with distinct nonzero values $\{c_1, c_2, \ldots, c_n\}$, let $E_i = \{x: \phi(x) = c_i\}$. Then

$$\phi = \sum_{i=1}^{n} c_i \chi_{E_i}.$$

Theorem 0.1.36. *Let $f: \mathbf{R}^1 \to \mathbf{R}^*$ be measurable. Then there exists a sequence $\{f_n\}$ of simple functions such that $f_n \to f$. If $f \geq 0$, the sequence $\{f_n\}$ can be constructed such that $0 \leq f_n$ and $f_n \leq f_{n+1}$ for each n.*

§0.2. The Lebesgue Integral

Definition 0.2.1. Let ϕ be a simple function which vanishes outside a set of finite measure. The *integral of* ϕ is defined by

$$\int \phi(x)\,dx = \sum_{i=1}^{n} a_i m A_i,$$

where $\{a_1, \ldots, a_n\}$ is the set of distinct nonzero values of ϕ and $A_i = \{x: \phi(x) = a_i\}$. We frequently use the abbreviated expression

$$\int \phi$$

for the integral. If E is any measurable set, then the *integral of* ϕ *over* E is defined by

$$\int_E \phi = \int \phi \chi_E.$$

If the a_i's in the range of ϕ are distinct, then any other description of ϕ as a linear combination of characteristic functions is simply a partitioning of the sets A_i. Since Lebesgue measure is additive, we see that the value of $\int \phi$ is independent of the description of ϕ in terms of characteristic functions.

Theorem 0.2.2. *Let* $E_1, E_2, \ldots, E_n$ *be disjoint measurable sets with*

$$E_0 = \bigcup_{k=1}^{n} E_k,$$

and let ϕ *be a simple function which is integrable on each set* E_k. *Then* ϕ *is integrable on* E_0 *and*

$$\int_{E_0} \phi = \sum_{k=1}^{n} \int_{E_k} \phi.$$

Theorem 0.2.3. *If* ϕ *and* ψ *are simple functions which are both integrable over a set* E, *and if* a *and* b *are real numbers, then* $a\phi + b\psi$ *is integrable over* E *and*

$$\int_E (a\phi + b\psi) = a \int_E \phi + b \int_E \psi.$$

Theorem 0.2.4. *If* ϕ *and* ψ *are simple functions,* ϕ *is integrable on* E *and for each* $x \in E$, $\phi(x) > |\psi(x)|$, *then* ψ *is integrable on* E *and*

$$\int_E \phi > \int_E \psi.$$

Theorem 0.2.5. *Let f be any non-negative, measurable function and E be any measurable set. If there exists a sequence $\{f_n\}$ of nondecreasing simple functions, each of which is integrable on E, such that for each $x \in E$,*

$$\lim_n f_n(x) = f(x)$$

and such that

$$\lim_n \int_E f_n < \infty,$$

we say that f is integrable on E *(or* summable over E*) and we define*

$$\int_E f = \lim_n \int_E f_n.$$

Definition 0.2.6. *If f is any measurable function, then f is integrable on E if and only if f^+ and f^- are both integrable on E. If this is the case, we define*

$$\int_E f = \int_E f^+ - \int_E f^-.$$

Theorem 0.2.7. *Let f and g be integrable on E. Then:*

(a) *For every real number c, the function cf is integrable on E and*

$$\int_E cf = c \int_E f.$$

(b) *The function $f + g$ is integrable on E and*

$$\int_E f + g = \int_E f + \int_E g.$$

(c) *If $f \leq g$ a.e., then*

$$\int_E f \leq \int_E g.$$

(d) *If A and B are disjoint measurable sets contained in E, then*

$$\int_{A \cup B} f = \int_A f + \int_B f.$$

(e) $|\int_E f| \leq \int_E |f|$.
(f) *If $f \geq 0$ a.e. on E and if $\int_E f = 0$, then $f = 0$ a.e. in E.*
(g) *If*

$$\int_A f = \int_A g$$

for every measurable $A \subset E$, then $f = g$ a.e. on E.

Again, it should be noted for $f + g$ is not defined at points where $f = \infty$ and $g = -\infty$ or where $f = -\infty$ and $g = \infty$. But the set of such points has measure zero since f and g are integrable. Hence, the integrability of $f + g$ and the value of the integral are independent of the choice of values for $f + g$ in these cases.

Theorem 0.2.8 (Fatou's Lemma). *If $\{f_n\}$ is a sequence of non-negative measurable functions and $f_n(x) \to f(x)$ a.e. on a set E, then*

$$\int_E f \le \varliminf \int_E f_n.$$

Theorem 0.2.9 (Lebesgue Monotone Convergence Theorem). *Let $\{f_n\}$ be an increasing sequence of non-negative measurable functions, each integrable on E, and let f be a function such that $\lim_n f_n = f$ a.e. Then f is integrable on E if and only if*

$$\lim_n \int_E f_n < \infty;$$

in this case, we have

$$\int_E f = \lim_n \int_E f_n.$$

Theorem 0.2.10 (Lebesgue Dominated Convergence Theorem). *Let g be integrable over E and let $\{f_n\}$ be a sequence of measurable functions such that $|f_n| \le g$ on E and $f_n(x) \to f(x)$ a.e. on E. Then*

$$\int_E f = \lim_n \int_E f_n.$$

Corollary 0.2.11. *Let $\{g_n\}$ be a sequence of integrable functions which converge a.e. to an integrable function g. Let $\{f_n\}$ be a sequence of measurable functions such that $|f_n| \le |g_n|$ and $f_n(x) \to f(x)$ a.e. If*

$$\int g = \lim_n \int g_n,$$

then

$$\int f = \lim_n \int f_n.$$

Theorem 0.2.12. *Let f be a non-negative function which is integrable over a set E. Then, for any $\varepsilon > 0$, there exists $\delta > 0$ such that for every set $A \subset E$ with* *mA < δ, we have*

$$\int_A f < \varepsilon.$$

§0.3. Vitali Covering Theorem

Definition 0.3.1. Let $E \subset \mathbf{R}^1$. Let V be a family of closed intervals (none consisting of a single point). The family V is said to cover E *in the sense of Vitali* if for every $x \in E$ and $\varepsilon > 0$, there exists a closed interval $v \in V$ such that $x \in v$ and $mv < \varepsilon$. In other words, E is covered by V in the sense of Vitali if every point $x \in E$ is in some arbitrarily small interval of V.

Theorem 0.3.2 (The Vitali Covering Theorem). *If the bounded set $E \subset \mathbf{R}^1$ is covered by a family V of closed intervals in the sense of Vitali, it is possible to find a finite or denumerable sequence $\{v_k\}$ of mutually disjoint intervals in V such that*

$$m^*\left[E - \bigcup_k v_k\right] = 0.$$

Proof [N1, p. 87]. Select any open interval Δ containing the set E and remove from V all the intervals which are not in Δ. Let V_0 denote those intervals remaining; then V_0 also covers E in the sense of Vitali. Let $v_1 \in V_0$. If $E \subset v_1$, then we have the desired result. If $E \not\subset v_1$, we will choose another interval from V_0 in the following manner. Suppose the mutually disjoint intervals v_1, $v_2, \ldots, v_n$ have already been chosen. If

$$E \subset \bigcup_{k=1}^{n} v_k,$$

then we are done. If, however,

(0.3.3) $$E - \bigcup_{k=}^{n} v_k \neq \varnothing,$$

let

$$F_n = \bigcup_{k=1}^{n} v_k,$$

$$G_n = \Delta - F_n.$$

Now consider those intervals in V_0 which are contained in the open set G_n. By (0.3.3), such intervals exist and their lengths are bounded by $m\Delta$. Let k_n denote the least upper bound of the lengths of these intervals and let v_{n+1} be any one of them for which

(0.3.4)
$$mv_{n+1} > \tfrac{1}{2}k_n.$$

Then v_{n+1} is disjoint from $v_1, v_2, \ldots, v_n$. If the process of constructing the intervals $\{v_k\}$ does not terminate after a finite number of steps (in which case the proof is complete), then we have constructed the desired sequence, i.e.,

(0.3.5)
$$m[E - S] = 0,$$

where

$$S = \bigcup_{k=1}^{\infty} v_k.$$

To prove (0.3.5), we construct for every k, a closed interval D_k having the same midpoint as the closed interval v_k but having five times greater length, i.e., $mD_k = 5mv_k$. Since the segments v_k are all disjoint and contained in the bounded set Δ, we have

(0.3.6)
$$\sum_{k=1}^{\infty} mv_k \le m\Delta;$$

therefore

(0.3.7)
$$\sum_{k=1}^{\infty} mD_k < \infty.$$

To prove (0.3.5), we will show that

(0.3.8)
$$E - S \subset \bigcup_{k=i}^{\infty} D_k$$

for all i. Let $x \in E - S$. Then $x \in G_i$ for all i. Since G_i is open, there exists an interval $v \in V_0$ such that $x \in v \subset G_i$. But v cannot be a subset of G_n for all n, since this would imply that

$$mv \le k_n < 2mv_{n+1}$$

for all n. This is impossible because $mv_n \to 0$ by (0.3.6). Therefore, for some n, $v \not\subset G_n$ and

(0.3.9)
$$v \cap F_n \ne \varnothing.$$

Let n be the smallest integer for which (0.3.9) holds. Since

$$v \cap F_i = \varnothing$$

and $F_1 \subset F_2 \subset \cdots$, then $n > i$. From the definition of n, we have

$$v \cap F_{n-1} = \varnothing.$$

Thus, we have

(0.3.10)
$$v \cap v_n \ne \varnothing,$$

$v \subset G_{n-1}$, and

(0.3.11)
$$mv \le k_{n-1} < 2mv_n.$$

From (0.3.10) and (0.3.11), it follows that $v \subset D_n$; therefore,

$$v \subset \bigcup_{k=i}^{\infty} D_k.$$

Thus,

$$x \in \bigcup_{k=i}^{\infty} D_k,$$

from which (0.3.8) follows. □

Corollary 0.3.12. *Under the hypotheses of the previous theorem, for every* $\varepsilon > 0$, *there exists a finite sequence* $v_1, v_2, \ldots, v_n$ *of mutually disjoint closed intervals in V for which*

$$m^* \left[E - \bigcup_{k=1}^{n} v_k \right] < \varepsilon.$$

One might suppose that the Vitali Covering Theorem would also hold for $\mathbf{R}^2$ by replacing the closed intervals of Theorem 0.3.2 by closed rectangles. However, a further condition must be added to prevent the rectangles from becoming "too slim," as seen in the following example attributed to Carathédory.

EXAMPLE 0.3.13 [J1, p. 112]. Let I denote the rectangle $\{(x, y) \in \mathbf{R}^2 : 0 < x < a, 0 < y < b\}$. Construct a sequence of rectangles I_p, $p = 1, 2, \ldots, m$, using the intervals $(0, a/m), (0, 2a/m), \ldots, (0, a)$ with altitudes $b, b/2, \ldots, b/m$, respectively. The area of each rectangle I_p is ab/m. Let

$$A_1 = \bigcup_{p=1}^{m} I_p.$$

Then

$$|A_1| = \frac{ab}{m}\left(1 + \frac{1}{2} + \cdots + \frac{1}{m}\right) = |I_p|\left(1 + \frac{1}{2} + \cdots + \frac{1}{m}\right).$$

Given $\eta > 0$, we can fix m so that

$$1 + \frac{1}{2} + \cdots + \frac{1}{m} > \frac{1}{\eta}$$

and

$$|I_p| < \eta |A_1|.$$

Through the points $a/m, 2a/m, \ldots, (m-1)a/m$ draw lines parallel to the y axis. These lines divide $I - A_1$ into $m - 1$ rectangles $J_2, J_3, \ldots, J_m$. On each of the rectangles J_i, $i = 2, \ldots, m$, determine sets A_i by the same method used in determining A_1 for I. Then

$$A_i \cap A_j = \varnothing, \quad i \neq j.$$

Let

$$\frac{1}{m}\left(1 + \frac{1}{2} + \cdots + \frac{1}{m}\right) = 1 - \theta_\eta.$$

Then $|A_1| = (1 - \theta_\eta)|I|$ and $|I - A_1| = \theta_\eta|I|$. Also,

$$|J_2 - A_2| = \theta_\eta|J_2|, \ldots, |J_m - A_m| = \theta_\eta|J_m|.$$

The set

$$R = \bigcup_{i=2}^{m} (J_i - A_i)$$

is the part of I not covered by the sets $A_1, A_2, \ldots, A_m$. Hence,

$$|R| = \theta_\eta \left| \bigcup_{i=2}^{m} J_i \right| = \theta_\eta|I - A_1| = \theta_\eta|I|.$$

If we repeat this process $k - 1$ times, we obtain $1 + (m - 1) + (m - 1)^2 + \cdots + (m - 1)^{k-1}$ mutually exclusive sets A_i and $(m - 1)^k$ rectangles J_r, where

$$\sum |J_r| < \theta_\eta^k|I|.$$

Any two rectangles in A_i have interior points in common. The rectangles contained in A_i together with those in J_r cover $\bar{I}$, the closure of I. Choose any mutually disjoint set of rectangles $\delta_1, \delta_2, \ldots, \delta_s$ from any of the sets A_i and J_r. Then no more than one of the δ_i can come from a single A_i and

$$|\delta_i| < \eta|A_i|.$$

If k is large enough so that $\theta_\eta^k < \eta$, then

(0.3.14) $$\left| \bigcup_{i=1}^{s} \delta_i \right| < \eta|I| + |J_r| < \eta|I| + \theta_\eta^k|I| < 2\eta|I|.$$

Now consider any finite set of such rectangles $\delta_1, \delta_2, \ldots, \delta_s$ which covers $\bar{I}$. If $P \in \bar{I}$, there is a first rectangle δ_P which contains P. Let $\delta(P)$ be the smallest rectangle which has P as its center and which contains δ_P. Then

(0.3.15) $$|\delta(P)| \leq 4|\delta_P|,$$

and the diameter[3] of $\delta(P)$ is not more than twice the diameter of I. If $\varepsilon > 0$ and $\eta > \varepsilon/8$, then by (0.3.14) and (0.3.15), each point $P \in I$ can be made the center of a rectangle $\delta(P)$ so that if $\delta(P_1)$, $\delta(P_2)$, $\ldots$ is any finite set of these

---

[3] The *diameter* of a set A, denoted diam(A), is defined as the least upper bound of $|a - b|$, $a, b \in A$.

rectangles which do not overlap, then

$$\sum |\delta(P_i)| \leq 4|\sum \delta_P| < \varepsilon |I|.$$

Let I be a rectangle and let $\rho > 0$. Divide I into K congruent rectangles I_k each with diameter less than $\frac{1}{2}\rho$. On each I_k, perform the preceding operations. Assign to each point $P \in \overline{I_k}$, a rectangle $\delta(P)$. Then the diameter of $\delta(P)$ is less than ρ. If $\delta(P_1)$, $\delta(P_2)$, ... is any finite nonoverlapping set of these rectangles, then

$$\sum |\delta(P_i)| < \varepsilon \sum |I_k| < \varepsilon |I|.$$

Perform this construction for $\rho = 1/2, 1/2^2, \ldots$, replacing ε by $\varepsilon/2, \varepsilon/2^2, \ldots$. Then the resulting set of rectangles $\delta(P)$ forms a covering of $\overline{I}$. If $\delta(P_1)$, $\delta(P_2)$, ... is any nonoverlapping set of these rectangles, then

$$\sum |\delta(P_i)| < \sum \frac{\varepsilon}{2^i}|I| < \varepsilon |I|.$$

Thus, the Vitali Covering Theorem fails for this covering; that is, it is not sufficient that the rectangles which form the covering have diameters tending to zero. An additional condition must be added; the ratio of the short side to the long side of the rectangles must be bounded away from zero. We now generalize such a condition of regularity for arbitrary coverings of sets in $\mathbf{R}^n$.

Definition 0.3.16 [S1, p. 57]. Given two points $a = (a_1, a_2, \ldots, a_n)$ and $b = (b_1, b_2, \ldots, b_n)$ in $\mathbf{R}^n$, the *closed interval* $[a, b] \in \mathbf{R}^n$ is the set

$$\{x = (x_1, x_2, \ldots, x_n) \in \mathbf{R}^n: a_i \leq x_i \leq b_i, i = 1, 2, \ldots, n\}.$$

If we replace the condition $a_i \leq x_i \leq b_i$ by the conditions $a_i < x_i < b_i$, $a_i \leq x_i < b_i$, $a_i < x_i \leq b_i$, we obtain the *open interval* (a, b), *the half-open (to the right) interval* $[a, b)$, and *the half-open (to the left) interval* $(a, b]$, respectively. A *cube* is an interval in which $b_1 - a_1 = b_2 - a_2 = \cdots = b_n - a_n$.

Definition 0.3.17 [S1, p. 106]. Let $E \subset \mathbf{R}^n$. The *parameter of regularity* of E, denoted $r(E)$, is the lower bound of the numbers

$$\frac{m^*(E)}{mJ}$$

where J is any cube containing E. When E is an interval,

$$\frac{l^n}{L^n} \leq r(E) \leq \frac{l}{L},$$

l is the smallest and L is the largest of the lengths of the sides of E. Thus, the parameter of regularity of a cube is 1.

Definition 0.3.18 [S1, p. 106]. A sequence of sets $\{E_k\}$ in $\mathbf{R}^n$ is called *regular* if there exists a positive number α such that $r(E_k) > \alpha$, $k = 1, 2, \ldots$.

Definition 0.3.19 [S1, p. 106]. A sequence of sets $\{E_k\}$ in $\mathbf{R}^n$ *converges* to a point $x \in \mathbf{R}^n$ if $x \in E_k$ for each k and

$$\lim_{k \to \infty} \operatorname{diam}(E_k) = 0.$$

Definition 0.3.20 [S1, p. 109]. Let $A \subset \mathbf{R}^n$. A family $\mathfrak{C}$ of sets covers A *in the sense of Vitali* if for every $x \in A$, there exists a regular sequence of sets from $\mathfrak{C}$ which converge to x.

Remark. We note that the covering constructed in Example 0.3.13 is not a covering in the sense of Vitali; it fails the regularity condition.

Theorem 0.3.21 (Vitali Covering Theorem, [S1, p. 109]). *If $A \subset \mathbf{R}^n$ and $\mathfrak{C}$ is a family of closed sets which covers A in the sense of Vitali, there exists a finite or denumerable sequence of disjoint sets $\{E_k\}$ in $\mathfrak{C}$ such that*

$$m^*\left(A - \bigcup_k E_k\right) = 0.$$

Proof. We will first prove the theorem under the following conditions:

(a) The parameters of regularity for all sets in $\mathfrak{C}$ are greater than a fixed number $\alpha > 0$.
(b) The set A is bounded, i.e., contained in an open sphere S.
(c) All the sets of $\mathfrak{C}$ are also contained in S.

The sequence $\{E_k\}$ will be defined by induction. For E_1, choose any set in $\mathfrak{C}$. Assume that the first p disjoint sets, $E_1, E_2, \ldots, E_p$, have been chosen. If $A \subset \bigcup_{i=1}^p E_i$, then $E_1, E_2, \ldots, E_p$ is the desired sequence. Otherwise, since $\mathfrak{C}$ is a Vitali covering, there must exist sets in $\mathfrak{C}$ which are disjoint from $\bigcup_{i=1}^p E_i$. Let δ_p denote the upper bound of the diameters of all the sets in $\mathfrak{C}$ which have no points in common with $\bigcup_{i=1}^p E_i$. Choose for E_{p+1} any one of these sets with diameter greater than $\frac{1}{2}\delta_p$. As noted, if the process terminates at some stage, the result is established. Therefore, we suppose that the process continues indefinitely. We will show that the sequence $\{E_k\}$ so defined covers almost all of A. Let

$$B = A - \bigcup_{k=1}^\infty E_k.$$

Suppose $m^*(B) > 0$. Because of assumption (a), we can associate with each set E_k a cube J_k such that $E_k \subset J_k$ and

$$\frac{m^*(E_k)}{m^*(J_k)} > \alpha.$$

Let $\overline{J_n}$ denote the cube with the same center as J_n and with diameter $(4n + 1) \operatorname{diam}(J_k)$. Because the sets E_k are disjoint, bounded, and have non-empty intersections with the bounded set A, then

$$\sum_{k=1}^{n} mE_k < \infty.$$

Thus, the series

$$\sum_k m(\overline{J_k}) = (4n + 1)^n \sum_k mJ_k \le (4n + 1)^n \alpha^{-1} \sum_k mE_k \le (4n + 1)^n \alpha^{-1} mS$$

converges. Therefore, we can find an integer K such that

$$\sum_{k=K+1}^{\infty} m\overline{J_k} < m^*(B).$$

[Moreover, $\lim_k \operatorname{diam}(\overline{J_k}) = 0$ and, thus, $\lim_k \operatorname{diam}(J_k) = 0$.] Hence, there exists a point $x_0 \in B$ which does not belong to any $\overline{J_k}$ for any $k > K$. By definition of B, x_0 does not belong to $\bigcup_{k=1}^{\infty} E_k$. Since the sets E_k are closed, there must exist a set $E \in \mathfrak{C}$ containing x_0, such that

(0.3.22) $E \cap E_k = \varnothing$ for $k = 1, 2, \ldots, K.$

Thus, the intersection of E and E_k must be nonempty for some $k > K$. For if not, we have

$$0 < \operatorname{diam}(E) \le \delta_k \le 2 \operatorname{diam}(E_{k+1}) \le 2 \operatorname{diam}(J_{k+1})$$

for every positive k. This contradicts the fact that $\lim_k \operatorname{diam}(J_k) = 0$. Let k_0 be the smallest value of k for which

$$E \cap E_k \ne \varnothing.$$

Then

(0.3.23) $E \cap E_k = \varnothing$ for $k = 1, 2, \ldots, k_0 - 1,$

giving

(0.3.24) $\operatorname{diam}(E) \le \delta_{k_0-1}.$

But, by (0.3.22), $k_0 > K$; this implies that $x_0 \notin \overline{J_{k_0}}$. Thus the set E contains points outside $\overline{J_{k_0}}$ and points in $E_{k_0} \subset J_{k_0}$. Hence,

$$\operatorname{diam}(E) > 2 \operatorname{diam}(J_{k_0}) \ge 2 \operatorname{diam}(E_{k_0}) > \delta_{k_0-1}.$$

This last statement contradicts (0.3.24). Therefore, our assumption that B has positive outer measure leads to a contradiction.

Now suppose $\mathfrak{C}$ is any family of closed sets which forms a Vitali covering of the set A. Let S_m denote the sphere centered at the origin of radius m. Let A_m denote those points x in $A \cap S_m$ for which there exists a sequence of sets in $\mathfrak{C}$ with parameters of regularity exceeding $1/m$ and converging to x. The sets A_m are an increasing sequence and $A = \lim_m A_m$. We can now define by induction a sequence of families of sets $\mathfrak{C}_n$ satisfying the following conditions:

(d) Each family $\mathfrak{C}_m$ consists of a finite number of disjoint sets in $\mathfrak{C}$, none of which for $m > 1$ have points in common with the preceding families $\mathfrak{C}_1$, $\mathfrak{C}_2, \ldots, \mathfrak{C}_{m-1}$.

(e) Letting C_m denote the union of all those sets in $\mathfrak{C}_m$,

$$\left| A_m - \bigcup_{i=1}^{m} C_i \right| < \frac{1}{m}.$$

To justify this last claim, suppose that for $m \le p$, the families $\mathfrak{C}_n$ have been chosen satisfying (d) and (e). Let

$$\overline{A_{p+1}} = A_{p+1} - \bigcup_{i=1}^{p} C_p.$$

Now consider the family of all sets in $\mathfrak{C}$ which are contained in the open set

$$\left(\bigcup_{i=1}^{p} C_i \right)^c$$

whose parameters of regularity exceed $1/(p + 1)$. This family of sets covers

$$\overline{A_{p+1}} \subset A_{p+1} \subset S_{p+1}$$

in the sense of Vitali. By the first part of this proof, we can extract from this family a sequence of disjoint sets $\{\overline{E_i}\}$ so as to cover $\overline{A_{p+1}}$ almost entirely. Therefore, for i_0 sufficiently large, we have

$$\left| A_{p+1} - \bigcup_{i=1}^{p} C_i - \bigcup_{i=1}^{i_0} \overline{E_i} \right| < \frac{1}{p + 1}.$$

Let $\mathfrak{C}_{p+1}$ be the family consisting of the sets $\overline{E_1}, \overline{E_2}, \ldots, \overline{E_{i_0}}$; then conditions (d) and (e) hold for $m = p + 1$. We now let $\mathfrak{C}' = \bigcup_m \mathfrak{C}_m$; this family consists of a denumerable number of disjoint sets from $\mathfrak{C}$; and if the union of these sets is denoted by C, we find

$$|A - C| = 0. \qquad \square$$

Remark. Saks attributes the preceding proof to Banach. Note that boundedness is no longer required as it was in Theorem 0.3.2. We state here some other versions of the theorem which are useful in applications. We omit the proofs, as they are similar to Theorem 0.3.21.

Theorem 0.3.25 [S1, p. 111]. *Given a set $A \subset \mathbf{R}^n$ and a family $\mathfrak{C}$ of closed sets, suppose that for each $x \in A$ there exists a number $\alpha > 0$, a sequence of sets $\{X_m\}$ in $\mathfrak{C}$, and a sequence of cubes $\{J_m\}$ such that $x \in J_m$, $X_m \subset J_m$, $|X_m| > \alpha |J_m|$ for $m = 1, 2, \ldots$, and $\lim_m \text{diam}(J_m) = 0$. Then $\mathfrak{C}$ contains a sequence of disjoint sets that covers A almost entirely.*

Theorem 0.3.26 [J2, p. 448]. *Assume $E \subset \mathbf{R}^n$ is a bounded set and $\mathscr{F}$ is a collection of open balls centered at points of E such that every point in E is the* *center of some ball in $\mathscr{F}$. Then there exists a sequence (possibly terminating) $B_1, B_2, \ldots$ of balls from $\mathscr{F}$ such that*

(a) *the balls $B_1, B_2, \ldots$ are disjoint,*
(b) $E \subset \bigcup_m 3B_m$;

that is, E is not covered by the disjoint balls but rather by the concentric balls of three times the radius.

§0.4. Baire Category Theorem and Baire Class Functions

Definition 0.4.1. A set E is *nowhere dense* if $(\bar{E})^c$ is dense, i.e., E contains no nonempty open subset. A set E is said to be of the *first category* (or *meager*) if E is the union of a countable collection of nowhere dense sets. A set which is not of the first category is said to be of the *second category* (or *nonmeager*) and the complement of a set of first category is called *residual* (or *comeager*).

Theorem 0.4.2 (Baire Category Theorem). *Let X be a complete metric space. Then no nonempty open subset of X is of the first category.*

Definition 0.4.3. Let Ω be a metric space. The class of Baire functions on Ω is the smallest class which contains the continuous finite, real-valued functions whose domains are Borel sets in Ω and which is closed under the operation of pointwise limits. The continuous functions are said to be of Baire class zero. The functions of Baire class n are those functions which are the pointwise limits of sequences of functions of class $n - 1$.

CHAPTER 1

Monotone Functions

Monotone functions play a very important role in the general theory of analysis of functions of a real variable. By virtue of their monotonic character, these functions have a wide variety of basic and elegant structural properties in terms of continuity, differentiability, and so on. In this chapter we establish some of the fundamental results in this direction.

§1.1. Continuity Properties

Definition 1.1.1. Let f be a real-valued finite function defined on a subset $S \subset \mathbf{R}^1$. Then f is said to be *increasing* (or *nondecreasing*) if, for every pair of points $x, y \in S$,

$$x < y \Rightarrow f(x) \le f(y).$$

If

$$x < y \Rightarrow f(x) < f(y),$$

then f is said to be *strictly increasing* on S. *Decreasing* functions are similarly defined. Functions which are either increasing or decreasing are called *monotone* (or *monotonic*).

We observe that if f is a decreasing function, then $-f$ is an increasing function. Thus, theorems analogous to those derived below for increasing functions may be established for decreasing functions.

Lemma 1.1.2. *If f is increasing on the closed interval $[a, b]$, then $f(c+)$ and*

$f(c-)$ both exist for each $c \in (a,b)$, and

$$f(c-) \leq f(c) \leq f(c+),$$

where

$$f(c+) = \lim_{x \to c+} f(x) \quad \text{and} \quad f(c-) = \lim_{x \to c-} f(x).$$

At the endpoints, we have $f(a) \leq f(a+)$ and $f(b-) \leq f(b)$.

Proof. Let c be a point in the open interval (a,b) and let $A = \{f(x): a < x < c\}$. Since f is increasing, A is bounded above by $f(c)$; thus, A has a supremum. Let $\alpha = \sup A$. By definition of supremum, for each $\varepsilon > 0$, there exists an $f(x_\varepsilon) \in A$ such that $f(x_\varepsilon) > \alpha - \varepsilon$. Since f is increasing, $f(x) > \alpha - \varepsilon$ if $x \in (x_\varepsilon, c)$. Hence,

$$\alpha - \varepsilon < f(x) \leq \alpha < \alpha + \varepsilon$$

if $x \in (x_\varepsilon, c)$; that is, $|f(x) - \alpha| < \varepsilon$ if $x \in (x_\varepsilon, c)$. Thus,

$$f(c-) = \lim_{x \to c-} f(x)$$

exists. The proofs for the existence of the right-hand limit of an interior point and the one-sided limits of the endpoints are similar. $\square$

Theorem 1.1.3. *If f is increasing on $[a,b]$, the set of discontinuities of f is countable.*

Proof. We first show that if $x_1, x_2, \ldots, x_n$ are arbitrary points in (a,b) such that $a = x_0 < x_1 < x_2 < \cdots < x_n < x_{n+1} = b$, then

$$f(a+) - f(a) + \sum_{k=1}^{n} [f(x_k+) - f(x_k-)] + f(b) - f(b-) \leq f(b) - f(a).$$

Assume that $y_k \in (x_k, x_{k+1})$. Then $f(x_k+) \leq f(y_k)$ and $f(y_{k-1}) \leq f(x_k-)$ so that

$$f(x_k+) - f(x_k-) \leq f(y_k) - f(y_{k-1}) \qquad k = 1, 2, \ldots, n.$$

Also

$$f(a+) - f(a) \leq f(y_0) - f(a)$$

and

$$f(b) - f(b-) \leq f(b) - f(y_n).$$

Adding together all of the above inequalities yields the desired result.

For each positive integer n, let $A_n = \{x \in (a,b): f(x+) - f(x-) \geq 1/n\}$. Then if $x_1, x_2, \ldots, x_k \in A_n$, we have

$$f(b) - f(a) \geq \sum_{i=1}^{k} [f(x_i+) - f(x_i-)] \geq \frac{k}{n},$$

i.e., k is finite. Thus, A_n is finite for each n; hence, $\bigcup_{n=1}^{\infty} A_n$ is countable. Since the set of discontinuities of f is equal to this union, the discontinuities of f must be countable. $\square$

Definition 1.1.4. Let f be a function defined on the interval $[a,b]$ and let $x \in (a,b)$. The numbers $f(x) - f(x-)$ and $f(x+) - f(x)$ are called the *left and right saltus*, respectively, of the function f at the point x. Their sum, $f(x+) - f(x-)$, is called the *saltus* (or *jump*) of the function f at this point. (For the points a and b, only one-sided jumps are considered.)

Definition 1.1.5. Let f be an increasing function defined on $[a,b]$. Let $\{x_1, x_2, \ldots, x_k, \ldots\}$ be the points of discontinuity of f. Define the function s:

$$s(a) = 0,$$

$$s(x) = f(a+) - f(a-) + \sum_{x_k < x} [f(x_k+) - f(x_k-)] + f(x) - f(x-),$$

$$a < x \le b.$$

The function s is called the *saltus function* (or *jump function*) of f. Note that s is an increasing function which remains constant on any open interval throughout which f is continuous.

Theorem 1.1.6. *The difference $\phi(x) = f(x) - s(x)$ between an increasing function and its saltus function is an increasing and continuous function.*

Proof. Let $a \le x < y \le b$. Then

$$s(y) - s(x) = \sum_{x < x_k < y} [f(x_k+) - f(x_k-)] + f(y) - f(y-) - f(x) + f(x-)$$

$$\le \sum_{x < x_k < y} [f(x_k+) - f(x_k-)] + f(y) - f(y-)$$

$$\le f(y) - f(x)$$

by Theorem 1.1.3. Hence,

$$f(x) - s(x) \le f(y) - s(y).$$

Thus, ϕ is increasing. If x is a point of continuity of f, then obviously

$$f(x+) - f(x) \le s(y) - s(x).$$

Assume x is not a point of continuity of f. Then

$$s(y) - s(x) = f(x+) - f(x-) + \sum_{x < x_k \le y} [f(x_k+) - f(x_k-)] + f(y) - f(y-)$$

$$\ge f(x+) - f(x),$$

i.e.,

$$f(x+) - f(x) \le s(y) - s(x) \quad \text{for } x < y.$$

Hence

$$f(x+) - f(x) \le \lim_{y \to x+} s(y) - s(x) = s(x+) - s(x-)$$

which implies $\phi(x+) \le \phi(x)$. Since ϕ is increasing, the reverse inequality also holds, yielding $\phi(x+) = \phi(x)$. In a similar manner, it can be proved that $\phi(x-) = \phi(x)$. Thus, ϕ is a continuous function. $\square$

§1.2. Differentiability Properties

In proving our principal result on the differentiability of monotone functions (Theorem 1.2.8), the concept of *derived numbers* will play an important role. We will consider derived numbers in more detail in Chapter 3.

Definition 1.2.1 [N1, p. 207]. The number λ (finite or infinite) is said to be a *derived number* of the function f at the point x_0 if there exists a sequence h_1, $h_2, h_3, \ldots (h_n \ne 0)$ such that $h_n \to 0$ and

$$\lim_{n \to \infty} \frac{f(x_0 + h_n) - f(x_0)}{h_n} = \lambda.$$

Symbolically, we say $\lambda = Df(x_0)$. If the (finite or infinite) derivative $f'(x_0)$ exists at the point x_0, then it will be a derived number $Df(x_0)$, and in this case, the function f will have no other derived numbers at the point x_0.

Lemma 1.2.2 [N1, p. 208]. *If the function f is defined on $[a,b]$, then derived numbers exist at every point $x \in [a,b]$.*

Proof. Let $x_0 \in [a,b]$ and let $\{h_n\}$ be a sequence such that $h_n \to 0$, $h_n \ne 0$, and $x_0 + h_n \in [a,b]$. Let $\sigma_n = [f(x_0 + h_n) - f(x_0)]/h_n$. If this sequence is bounded, then by the Bolzano–Weierstrass theorem, it has at least one limit point, call it λ. Therefore, $\{\sigma_n\}$ has a subsequence $\{\sigma_{n_k}\}$ such that $\sigma_{n_k} \to \lambda$. Then $\lambda = Df(x_0)$. If $\{\sigma_n\}$ is unbounded above, we can extract a subsequence $\{\sigma_{n_k}\}$ such that $\sigma_{n_k} \to \infty$. Then $Df(x_0) = \infty$. $\square$

Lemma 1.2.3 [N1, p. 208]. *If the function f is increasing on $[a,b]$, then all of its derived numbers are non-negative.*

Proof. The proof is immediate since f increasing implies that $f(x_0 + h_n) - f(x_0)$ has the same sign as h_n. $\square$

Lemma 1.2.4 [N1, p. 208]. *Let f be a strictly increasing function defined on $[a,b]$ and let $p \ge 0$. If at every point x of the set $E \subset [a,b]$, there exists at least one derived number $Df(x)$ such that*

$$Df(x) \le p,$$

then

$$m^*(f(E)) \leq p \cdot m^*(E).$$

Proof. Let $\varepsilon > 0$ and let G be a bounded open set containing E such that $mG < m^*(E) + \varepsilon$. Let p_0 be any number such that $p_0 > p$. If $x_0 \in E$, there exists a sequence $\{h_n\}$ such that $h_n \to 0$ and

$$\lim_{n \to \infty} \frac{f(x_0 + h_n) - f(x_0)}{h_n} = Df(x_0) \leq p < p_0.$$

For sufficiently large n, the closed interval $[x_0, x_0 + h_n]$ will be contained in G (assuming $h_n > 0$, otherwise use the interval $[x_0 + h_n, x_0]$). Also

$$\frac{f(x_0 + h_n) - f(x_0)}{h_n} < p_0$$

for sufficiently large n. Now consider the closed intervals

$$I_n(x_0) = [x_0, x_0 + h_n],$$

$$\Delta_n(x_0) = [f(x_0), f(x_0 + h_n)].$$

Since f is strictly increasing, $f[I_n(x_0)] \subset \Delta_n(x_0)$. The lengths of these intervals are

$$mI_n(x_0) = |h_n|,$$

$$m\Delta_n(x_0) = |f(x_0 + h_n) - f(x_0)| < p_0 \cdot |h_n| = p_0 \cdot mI_n(x_0).$$

Since $h_n \to 0$, there must exist an arbitrarily small interval among the $\Delta_n(x_0)$. Now $f(E)$ is covered by $\{\Delta_n(x): x \in E\}$ in the sense of Vitali. Thus, we can select a countable sequence of pairwise disjoint intervals $\{\Delta_{n_i}(x_i)\}$ such that

$$m\left(f(E) - \bigcup_{i=1}^{\infty} \Delta_{n_i}(x_i) \right) = 0.$$

Now

$$m^*(f(E)) \leq \sum_{i=1}^{\infty} m\Delta_{n_i}(x_i) < p_0 \sum_{i=1}^{\infty} mI_{n_i}(x_i).$$

Since f is strictly increasing (and hence one-to-one) and $\Delta_{n_i}(x_i)$ are pairwise disjoint, then $I_{n_i}(x_i)$ are also pairwise disjoint. Therefore,

$$\sum_{i=1}^{\infty} mI_{n_i}(x_i) = m\left(\bigcup_{i=1}^{\infty} I_{n_i}(x_i) \right).$$

Since $\bigcup_{i=1}^{\infty} I_{n_i}(x_i) \subset G$, we have

$$m^*(f(E)) < p_0 \cdot mG < p_0 \cdot (m^*(E) + \varepsilon).$$

Letting $p_0 \to p$ and $\varepsilon \to 0$, we get $m^*(f(E)) \leq p \cdot m^*(E)$. $\qquad\square$

Proceeding analogously to the above lemma we also have

Lemma 1.2.5 [N1, p. 210]. *Let f be a strictly increasing function defined on* $[a, b]$. *If* $E \subset [a, b]$ *and, if at every point* $x \in E$, *there exists at least one derived number* $Df(x)$ *such that*

$$Df(x) \geq q \quad (q \geq 0),$$

then

$$m^*(f(E)) \geq q \cdot m^*(E).$$

Theorem 1.2.6 [N1, p. 211]. *The set of points at which at least one derived number of an increasing function f is infinite is of measure zero.*

Proof. Let $E = \{x \in [a, b]: Df(x) = +\infty\}$. Suppose f is strictly increasing and $m^*(E) > 0$. Then by the previous result, $m^*(f(E)) = \infty$. But $f(E) \subset [f(a), f(b)]$. Hence, $f(E)$ cannot have infinite measure. Therefore, $m^*(E)$ must be zero. Now if f is not strictly increasing, consider the function

$$g(x) = f(x) + x.$$

Then g is strictly increasing and

$$\frac{g(x + h) - g(x)}{h} = \frac{f(x + h) + (x + h) - f(x) - x}{h}$$

$$= \frac{f(x + h) - f(x)}{h} + 1.$$

Therefore, the set of points where $Df(x)$ is infinite is the same set of points where $Dg(x)$ is infinite and thus has measure zero. $\square$

Theorem 1.2.7 [N1, p. 211]. *Let f be an increasing function defined on the closed interval* $[a, b]$ *and let p and q be two numbers such that* $p < q$. *If at every point of the set* $E_{p,q} \subset [a, b]$, *there exist two derived numbers* $D_1 f(x)$ *and* $D_2 f(x)$ *such that* $D_1 f(x) < p < q < D_2 f(x)$, *then* $mE_{p,q} = 0$.

Proof. First, suppose f is strictly increasing and use Lemmas 1.2.4 and 1.2.5 to get $m^*(f(E_{p,q})) \leq p \cdot m^*(E_{p,q})$ and $m^*(f(E_{p,q})) \geq q \cdot m^*(E_{p,q})$. Thus,

$$q \cdot m^*(E_{p,q}) \leq m^*(f(E_{p,q})) \leq p \cdot m^*(E_{p,q}).$$

Since $p < q$, the above can hold only if $m^*(E_{p,q}) = 0$.

If f is not strictly increasing, consider the function

$$g(x) = f(x) + x.$$

Since

$$\frac{g(x + h) - g(x)}{h} = \frac{f(x + h) - f(x)}{h} + 1,$$

then for any sequence $\{h_n\}$ such that $h_n \neq 0$ and $h_n \to 0$, we have

$$\lim_{n \to \infty} \frac{g(x + h_n) - g(x)}{h_n} = \lim_{n \to \infty} \frac{f(x + h_n) - f(x)}{h_n} + 1.$$

Therefore, $Dg(x) = Df(x) + 1$. Then

$$D_1 f(x) < p < q < D_2 f(x)$$
$$\Rightarrow D_1 g(x) < p + 1 < q + 1 < D_2 g(x).$$

Hence, by the first part of this theorem, $m^*(E_{p,q}) = 0$. $\square$

Theorem 1.2.8 [N1, p. 211]. *Every increasing function f defined on a closed interval $[a, b]$ has a finite derivatives a.e. on $[a, b]$.*

Proof. Let $E = \{x \in [a, b]: f'(x)$ does not exist$\}$. If $x_0 \in E$, there are two distinct derived numbers $D_1 f(x_0)$ and $D_2 f(x_0)$. Suppose that $D_1 f(x_0) < D_2 f(x_0)$. Then we can find rational numbers p and q such that

$$D_1 f(x_0) < p < q < D_2 f(x_0).$$

Thus,

$$E = \bigcup_{p,q \in Q} E_{p,q}$$

has measure zero. Hence, $f'(x)$ exists almost everywhere, and since $f'(x)$ is infinite only on a set of measure zero (Theorem 1.2.6), the conclusion follows. $\square$

Remark. Another interesting proof, which is geometric in nature, may be seen in the literature. The proof uses the "rising sun lemma" [T4, pp. 61–64].

§1.3. Reconstruction of f from f'

Theorem 1.3.1. *If f is an increasing function defined on $[a, b]$, then its derivative f' is measurable and*

$$(1.3.2) \qquad \int_a^b f'(x)\, dx \leq f(b) - f(a)$$

so that f' is summable.

Proof. Extend the definition of f so that

$$f(x) = f(b), \quad b < x \leq b + 1.$$

Let

$$g(x) = \lim_{n \to 0} \frac{f(x + h) - f(x)}{h}.$$

Then g is defined almost everywhere and f is differentiable wherever g is finite. Let

$$g_n(x) = n\left[f\left(x + \frac{1}{n}\right) - f(x)\right].$$

Then $g_n(x) \to g(x)$ for almost all x. Since $f(x + 1/n)$ and $f(x)$ are increasing, they are measurable. Therefore, each $g_n(x)$ is measurable; hence, g, the limit of an almost everywhere convergent sequence of measurable functions, is measurable. Since $f'(x) = g(x)$ almost everywhere, f' is measurable. By Fatou's lemma

$$\int_a^b g \le \underline{\lim} \int_a^b g_n = \underline{\lim} n \int_a^b \left[f\left(x + \frac{1}{n}\right) - f(x)\right].$$

Since f is monotonic, $\int_a^b f(x + 1/n)$ can be considered a Riemann integral. Changing variables

$$\int_a^b f\left(x + \frac{1}{n}\right) dx = \int_{a+1/n}^{b+1/n} f(x)\, dx.$$

Then

$$\int_a^b \left[f\left(x + \frac{1}{n}\right) - f(x)\right] dx = \int_a^b f\left(x + \frac{1}{n}\right) dx - \int_a^b f(x)\, dx$$

$$= \int_{a+1/n}^{b+1/n} f(x)\, dx - \int_a^b f(x)\, dx$$

$$= \int_{a+1/n}^{b} f(x)\, dx + \int_b^{b+1/n} f(x)\, dx - \int_a^{a+1/n} f(x)\, dx$$

$$\quad - \int_{a+1/n}^{b} f(x)\, dx$$

$$= \int_b^{b+1/n} f(x)\, dx - \int_a^{a+1/n} f(x)\, dx$$

$$= \frac{1}{n} \cdot f(b) - \int_a^{a+1/n} f(x)\, dx.$$

Now $f(x) \ge f(a)$ if $x \in [a, a + 1/n]$. Hence,

$$-\int_a^{a+1/n} f(x)\, dx \le -\int_a^{a+1/n} f(a)\, dx = -\frac{1}{n} \cdot f(a).$$

Therefore,

$$\int_a^b g \le \underline{\lim} \left\{ n \cdot \int_a^b \left[f\left(x + \frac{1}{n}\right) + f(x)\right] dx \right\}$$

$$= \underline{\lim} n \cdot \left(\frac{1}{n} \cdot f(b) - \int_a^{a+1/n} f(x)\, dx \right)$$

$$= \varliminf \left[f(b) - n \cdot \int_a^{a+1/n} f(x)\, dx \right]$$

$$\leq \varliminf \left(f(b) - n \cdot \frac{1}{n} \cdot f(a) \right) = f(b) - f(a). \qquad \square$$

Remark. In elementary calculus, we learn the familiar formula

$$\int_a^b f'(x)\, dx = f(b) - f(a).$$

However, as seen in the following examples, the conditions of Theorem 1.3.1 do not always yield equality in (1.3.2).

EXAMPLE 1.3.3. First, we construct the *Cantor ternary set*. Let $I = [0, 1]$. Divide I into three equal parts and remove the interior of the middle third interval:

$$E_{11} = \left(\frac{1}{3}, \frac{2}{3} \right).$$

Divide the two remaining closed intervals

$$\left[0, \frac{1}{3} \right], \qquad \left[\frac{2}{3}, 1 \right]$$

into three equal parts and remove the interior of each of the middle parts:

$$E_{21} = \left(\frac{1}{3^2}, \frac{2}{3^2} \right), \qquad E_{22} = \left(\frac{7}{3^2}, \frac{8}{3^2} \right).$$

Next we divide each of the remaining intervals into thirds and remove the interior of each of the middle parts, so that

$$E_{31} = \left(\frac{1}{3^3}, \frac{2}{3^3} \right), \qquad E_{32} = \left(\frac{7}{3^3}, \frac{8}{3^3} \right), \qquad E_{33} = \left(\frac{19}{3^3}, \frac{20}{3^3} \right),$$

$$E_{34} = \left(\frac{25}{3^3}, \frac{26}{3^3} \right).$$

Continuing in this manner, the 2^{n-1} open intervals removed at the nth step will be E_{ni} $(i = 1, 2, \ldots, 2^{n-1})$. The length of E_{ni} will be 3^{-n}. The Cantor ternary set C is

$$C = I - \bigcup_{n=1}^{\infty} E_n,$$

where

$$E_n = \bigcup_{i=1}^{2^{n-1}} E_{ni}.$$

To better understand the Cantor set, we introduce binary and ternary notation. Thus, in ternary notation $._3101 = \frac{1}{3} + \frac{1}{27} = \frac{10}{27}$, and in binary notation $._2101 = \frac{1}{2} + \frac{1}{8} = \frac{5}{8}$. Every $x \in I$ has a ternary representation of the form

$$._3a_1a_2a_3\ldots,$$

where each a_n is either 0, 1, or 2; and each such number is in $[0, 1]$. However, representation of this type is not unique. For example, the number $\frac{1}{3} = ._31 = ._302222\ldots$.

Now let us consider the first interval removed from I in the construction of the Cantor set, namely $E_{11} = (\frac{1}{3}, \frac{2}{3})$. If $x \in E_{11}$ and $x = ._3a_1a_2a_3\ldots$, then a_1 must be 1. Each of the endpoints of E_{11} has two representations:

$$\frac{1}{3} = ._3100\ldots = ._30222\ldots,$$

$$\frac{2}{3} = ._31222\ldots = ._32000\ldots.$$

No remaining point of I can have a 1 in the first position of its ternary representation. Thus, at the first stage of construction of C, we remove all those points $x = ._3a_1a_2a_3\ldots$, where a_1 must be 1, and only those points are removed.

Similarly, in the second step of construction, we remove those points such that a_2 must be 1, and only those points are removed. Thus, at the nth step of construction, we remove those points and only those points such that a_n must be 1. Then C must consist of those points which have ternary representations that contain only the digits 0 or 2.

It is clear that the Cantor set contains the endpoints of the intervals E_{ni}, a countable set. We list the endpoints of some of the intervals:

$$E_{11}: \quad \frac{1}{3} = ._31 = ._302222\ldots; \quad \frac{2}{3} = ._3122\ldots = ._32$$

$$E_{21}: \quad \frac{1}{9} = ._301 = ._30022\ldots; \quad \frac{2}{9} = ._30122\ldots = ._302$$

$$E_{22}: \quad \frac{7}{9} = ._321 = ._32022\ldots; \quad \frac{8}{9} = ._32122\ldots = ._322$$

$$E_{31}: \quad \frac{1}{27} = ._3001 = ._300022\ldots; \quad \frac{2}{27} = ._300122\ldots = ._30002$$

$$E_{32}: \quad \frac{7}{27} = ._3021 = ._302022\ldots; \quad \frac{8}{27} = ._302122\ldots = ._3022$$

$$E_{33}: \quad \frac{19}{27} = ._3201 = ._320022\ldots; \quad \frac{20}{27} = ._320122\ldots = ._3202$$

$$E_{34}: \quad \frac{25}{27} = ._3221 = ._322022\ldots; \quad \frac{26}{27} = ._322122\ldots = ._3222$$

We now see that the endpoints of the intervals E_{ni} have ternary representations where all the digits after a certain place are either 0 or 2. Thus, there are points in C which are not the endpoints of any interval E_{ni}, i.e., those points whose ternary representations contain an infinite number of 0's and an infinite number of 2's.

The reader should now verify the following observations [T2, pp. 86–88]:

(1) C is a closed set.
(2) C contains no interior points.
(3) Every point of C is an accumulation point of C.
(4) C is an uncountably infinite set of measure zero.

Now we will construct a function ω, often referred to as the *Cantor ternary function* or the *Cantor function* [HT]. We will let $a = 0$ or 2 and $b = a/2$ so that b assumes only the values 0 or 1. If $x = .{}_3 a_1 a_2 a_3 \ldots \in C$, we define

$$\omega(x) = .{}_2 b_1 b_2 b_3 \ldots.$$

According to this definition, ω has equal values

$$\begin{aligned}
\omega_{ni} &= \omega(.{}_3 a_1 a_2 \ldots a_m 0222 \ldots) \\
&= .{}_2 b_1 b_2 \ldots b_m 0111 \ldots \\
&= .{}_2 b_1 b_2 \ldots b_m 1000 \ldots \\
&= \omega(.{}_3 a_1 a_2 \ldots a_m 2000 \ldots) \\
&= \frac{(2i - 1)}{2^n}
\end{aligned}$$

at the endpoints of each interval E_{ni} and we take this value as the value for ω on the entire interval E_{ni}.

First, we will show that ω is an increasing function. In proving the inequality

$$\omega(x') \le \omega(x'') \quad \text{if } x' < x'',$$

we may assume that the points under consideration are in C, since ω is constant on each interval E_{ni}. Let

$$x' = .{}_3 a_1' a_2' \ldots,$$

$$x'' = .{}_3 a_1'' a_2'' \ldots.$$

If $x' < x''$, there will be a value of the subscript p such that

$$a_1' = a_1'', \ldots, a_{p-1}' = a_{p-1}'',$$

but

$$a_p' < a_p''.$$

Hence,

$$\omega(x') = .{}_2 b_1' b_2' \ldots \le .{}_2 b_1'' b_2'' \ldots = \omega(x'').$$

Next we will show that ω is continuous by showing that $\omega(x') \to \omega(x)$ as $x' \to x$. Again, it suffices to consider only the case where $x \in C$ and x' takes on values in C only. Consider the case of $x' > x$. Let

$$x = {}_{.3}a_1a_2\ldots,$$

$$x' = {}_{.3}a_1'a_2'\ldots.$$

If $x < x'$ and $x' \to x$, then there will be a value of the subscript p (where $p \to \infty$ as $x' \to x$) such that

$$a_1' = a_1, \ldots, a_{p-1}' = a_{p-1}$$

but $a_p < a_p'$. Hence,

$$\omega(x') = {}_{.2}b_1b_2\ldots b_{p-1}b_p'\ldots \to {}_{.2}b_1b_2\ldots b_{p-1}b_p\ldots = \omega(x).$$

The case $x' < x$ is similarly proved.

Alternatively, we could have proved continuity by observing that the range of ω is dense in $[0, 1]$.

It is clear that the derivative $\omega' = 0$ at all points of the set $I - C$, that is, almost everywhere. Thus,

$$\int_0^1 \omega'(x)\,dx = 0 \neq \omega(1) - \omega(0) = 1.$$

In the exercises and in Chapter 8, we will generalize the concept of the Cantor ternary set and Cantor function.

EXAMPLE 1.3.4 [J1, p. 146]. We remark here that monotone functions are not the only ones for which

$$\int_a^b f'(x)\,dx = f(b) - f(a)$$

may not hold. We present an example in which the equality fails for the reason that f' is not summable. Let f be defined on $[0, 1]$ by

$$f(x) = \begin{cases} 0, & x = 0 \\ x^2 \sin \dfrac{1}{x^2}, & x \neq 0. \end{cases}$$

For $x \neq 0$,

$$f'(x) = 2x \sin \frac{1}{x^2} - \frac{2}{x} \cos \frac{1}{x^2}.$$

When $x = 0$,

$$\lim_{h \to 0} \frac{f(0 + h) - f(0)}{h} = \lim_{h \to 0} h \sin \frac{1}{h^2} = 0.$$

Hence, $f'(0) = 0$, $f'(x)$ is finite at each $x \in [0, 1]$, and f' is bounded on every closed interval not containing the origin. However, f' is not summable. For $n \geq 1$, let

$$x_n' = \sqrt{\frac{2}{(4n-3)\pi}}.$$

Then $\sin(1/x_n'^2) = 1$. There is a first point x_n to the left of x_n' such that $f(x_n) = 0$ and $x_n > x_{n+1}'$. On the interval $[x_n, x_n']$, f' is bounded; thus,

$$\int_{x_n}^{x_n'} f'(x)\,dx = f(x_n') - f(x_n) = \frac{2}{(4n-3)\pi}.$$

Then

$$\int_{x_n}^{1} |f'(x)|\,dx \geq \sum_{i=1}^{n} \frac{2}{(4i-3)\pi}.$$

As $n \to \infty$, the right-hand side becomes infinite. Thus, $|f'|$ is not summable; therefore, f' is not summable.

§1.4. Series of Monotone Functions

The next result we present is due to Fubini; we shall hereafter refer to this theorem as Fubini's theorem although there is another more well-known result concerning iterated integrals which is also called Fubini's theorem.

Theorem 1.4.1. *If $\{f_n\}$, $n = 1, 2, \ldots$, is a sequence of increasing functions defined on $[a, b]$, and if the series*

$$\sum_{n=1}^{\infty} f_n(x)$$

converges to a sum function $s(x)$ at each point of $[a, b]$, then

$$s'(x) = \sum_{n=1}^{\infty} f_n'(x)$$

for almost all $x \in [a, b]$.

Proof. Let E be the set of points $x \in (a, b)$ such that the derivatives $s'(x)$, $f_1'(x)$, $f_2'(x)$, ... all exist and are finite. Then $[a, b] - E$ has measure zero. Let

$$s_n(x) = \sum_{k=1}^{n} f_k(x).$$

Suppose $x \in E$ and $h > 0$ is sufficiently small such that $x + h \in [a, b]$. Then

$$\frac{s(x+h) - s(x)}{h} = \sum_{k=1}^{\infty} \frac{f_k(x+h) - f_k(x)}{h}.$$

For any n,

$$\frac{s(x+h) - s(x)}{h} \geq \sum_{k=1}^{n} \frac{f_k(x+h) - f_k(x)}{h}.$$

Letting $h \to 0$, we get

$$s'(x) \geq \sum_{k=1}^{n} f_k'(x).$$

The functions f_k' are non-negative and the sequence

$$\sum_{k=1}^{n} f_k'(x)$$

is bounded above by $s'(x)$. Hence, this sequence converges a.e. on $[a, b]$. Now we will show its limit is $s'(x)$. Since

$$\lim_{n \to \infty} s_n(b) = s(b),$$

there exists a subsequence $\{s_{n_j}(b)\}$ such that

$$0 \leq s(b) - s_{n_j}(b) \leq 2^{-j}.$$

Because $s - s_{n_j}$ is an increasing function, we have

$$0 \leq s(x) - s_{n_j}(x) \leq 2^{-j}$$

for all $x \in [a, b]$. Therefore, the series

$$\sum_{j=1}^{\infty} s - s_{n_j}$$

is a convergent series of increasing functions. By the same reasoning we applied above to the convergent series of increasing functions $\{f_n\}$, we conclude that the series

$$\sum_{j=1}^{\infty} s' - s_{n_j}'$$

is convergent a.e. in $[a, b]$. Hence, $s' - s_{n_j}' \to 0$ as $j \to \infty$. But the sequence $\{s_n'\}$ is nondecreasing in n. Thus, $s'(x) - s_n'(x) \to 0$ as $n \to \infty$ for almost all $x \in [a, b]$. $\square$

Theorem 1.4.2 [BH1]. *Let $\{f_n\}$ be a sequence of increasing (or decreasing) functions which converge at each point of the interval $[a, b]$ to a continuous function f. Then f is increasing (or decreasing) and $f_n \to f$ uniformly on $[a, b]$.*

Proof. We first prove that if $\{f_n\}$ is a sequence of increasing functions, then f is increasing. Suppose that this is not the case. Then there exists a pair of points x_1, x_2 ($x_1 < x_2$) in $[a, b]$ such that $f(x_1) > f(x_2)$. Let $h = f(x_1) -$

$f(x_2)$ and $0 < \varepsilon < h/2$. Since $f_n(x_1) \to f(x_1)$, there exists $N(\varepsilon, x_1)$ such that

$$|f_n(x_1) - f(x_1)| \le \varepsilon \quad \text{for } n \ge N(\varepsilon, x_1).$$

Similarly, there exists $N(\varepsilon, x_2)$ such that

$$|f_n(x_2) - f(x_2)| \le \varepsilon \quad \text{for } n \ge N(\varepsilon, x_2).$$

Let $N = \max(N(\varepsilon, x_1), N(\varepsilon, x_2))$. Then for $n \ge N$,

$$f_n(x_2) - f_n(x_1) < f(x_2) + \frac{h}{2} - f(x_1) + \frac{h}{2} = h - (f_1(x_1)) - f(x_2) = h - h = 0;$$

that is, $f_n(x_1) - f_n(x_2) > 0$. Since $x_1 < x_2$, this contradicts the fact that f_n is increasing. Thus, the assumption that f is not increasing is false.

Now we prove that $f_n \to f$ uniformly on $[a, b]$. For $\varepsilon > 0$, let $x_1, x_2, \ldots, x_m$ be a finite set of points such that $a = x_1 < x_2 < \cdots < x_m = b$ and

$$0 \le f(x_k) - f(x_{k-1}) \le \varepsilon \quad \text{for } k = 2, 3, \ldots, m.$$

Since f is uniformly continuous and increasing on $[a, b]$, this is possible by making the distance between consecutive x_k's small. Since f is increasing, we have

$$0 \le f(x_k) - f(x) \le \varepsilon \quad \text{for } x_{k-1} \le x \le x_k.$$

Since f_n converges pointwise, we can choose n large enough so that

$$|f_n(x_k) - f(x_k)| \le \varepsilon \quad \text{for } k = 1, 2, \ldots, m.$$

Each $x \in [a, b]$ is in some interval $[x_{k-1}, x_k]$; therefore,

$$f_n(x) \le f_n(x_k) \le f(x_k) + \varepsilon \le f(x) + 2\varepsilon.$$

Similarly,

$$f(x) - 2\varepsilon < f(x_{k-1}) \le f_n(x_{k-1}) \le f_n(x).$$

Combining these last two inequalities, we get

$$|f_n(x) - f(x)| \le 2\varepsilon$$

for every $x \in [a, b]$ if n is large enough. Thus, we have uniform convergence. $\qquad\square$

EXERCISES

1. Let $x \in [0, 1]$ be expressed as a finite or infinite decimal expansion $x = .a_1 a_2 \ldots a_n \ldots$, and let

$$f(x) = \left(\frac{a_1}{10}\right)^2 + \left(\frac{a_2}{100}\right)^2 + \cdots.$$

Show that f is monotone and is discontinuous at every value of x represented by a finite decimal. The function

$$f(x) = .0a_1 0a_2 0a_3 \ldots$$

has similar properties.

2. From Theorem 1.1.3, we know that there can be at most a countably infinite number of jumps of a monotone function. But this problem illustrates that the set of discontinuities can be as "nasty" as possible, even everywhere dense. Let $x_1, x_2,$ ... be an everywhere dense denumerable set in (a, b). Let a function $u(x) > 0$ be assigned to each point x_i of this set in such a way that $\sum_{i=1}^{\infty} u(x_i)$ converges. Let E be a measurable subset of (a, b), and let

$$\phi(x) = m[E \cap (a, x)] + \sum_{x_i \in (a, x)} u(x_i).$$

Show that ϕ is increasing on $[a, b]$ and that the discontinuities of ϕ are everywhere dense on $[a, b]$. Show that $\phi'(x) = 1$ at almost all points of E, and that $\phi'(x) = 0$ at almost all points of $[a, b] - E$.

3. [J2, pp. 521–527] **Definition.** An *elementary jump function* is a function σ defined on $\mathbf{R}^1$ with the following form:

$$\sigma(x) = \begin{cases} A, & x < t \\ B, & x = t \\ C, & x > t \end{cases}$$

with $A \leq B \leq C$ and $A < C$. Let $I \subset \mathbf{R}^1$ be a compact interval. An *increasing jump function* on I is a function $s: I \to \mathbf{R}^1$ such that, for $x \in I$,

$$s(x) = \sum_{k=1}^{\infty} \sigma_k(x),$$

where $\sigma_1, \sigma_2, \ldots$ are elementary increasing jump functions.

Show that the saltus function of an increasing function f, as given by Definition 1.1.5, is an increasing jump function. For this reason, increasing jump functions are also called saltus-type functions.

4. Theorem 1.1.6 gives one representation of an increasing function f as the sum of a continuous function and a jump function. Prove that the representation of f in the form

$$f(x) = \phi(x) + s(x),$$

where ϕ is continuous and s is a jump function is unique except for the addition of constants to ϕ and s.

5. (Generalized Cantor sets) Let $E_0 = [0, 1]$ and let $\{\delta_n\}$ be a strictly decreasing sequence such that $\delta_0 = 1$ and $\delta_n \to 0$. Suppose that E_n is the union of 2^n disjoint closed intervals contained in E_0, each interval having length $2^{-n}\delta_n$. From each of the component intervals of E_n, delete a segment in the center such that the remaining 2^{n+1} intervals each has length $2^{-n-1}\delta_{n+1}$. (Note $\delta_{n+1} < \delta_n$.) Let E_{n+1} be the union of these 2^{n+1} intervals. Then $E_1 \supset E_2 \supset \cdots$ and $m(E_n) = \delta_n$. Let $E = \bigcap_{n=0}^{\infty} E_n$. Show E is compact, $m(E) = 0$, and E is perfect. Observe if $\delta_n = (\frac{2}{3})^n$, then we have the Cantor ternary set.

6. (Continuation of problem 5). Let

$$g_n = \frac{\chi_{E_n}}{\delta_n} \quad \text{and} \quad f_n(x) = \int_0^x g_n(t)\, dt.$$

Then, for each n, $f_n(0) = 0$, $f_n(1) = 1$, f_n is monotone, and f_n is constant on each segment in E_n^C. If I is one of the intervals contained in E_n, then

$$\int_I g_n = \int_I g_{n+1} = 2^{-n}.$$

Thus, $f_{n+1}(x) = f_n(x)$, $x \notin E_n$. Show that $\{f_n\}$ is uniformly convergent to a continuous monotone function f such that $f(0) = 0$, $f(1) = 1$, and $f'(x) = 0$ for $x \notin E$, i.e., almost everywhere. Thus,

$$f(x) - f(a) = \int_a^x f'(t)\,dt$$

fails to hold.

7. If f is increasing on an interval $[a, b]$ and x_0 is a point at which f is discontinuous, show that

$$\lim_{x \to x_0-} f(x) = f(x_0-) < f(x_0+) = \lim_{x \to x_0+} f(x).$$

Thus, corresponding to x_0, there exists a rational number in the interval $(f(x_0-), f(x_0+))$. Use this to show that the set of points of discontinuity of f is at most countable.

8. Let $S = \{a_n\}$, $a_n \in \mathbf{R}^1$. Let $\{d_n\}$ be a sequence of positive numbers such that

$$\sum_{n=1}^{\infty} d_n < \infty.$$

Define

$$\phi_{a_n}(x) = \begin{cases} 0, & x \le a_n \\ 1, & x \ge a_n. \end{cases}$$

Show

$$f(x) = \sum_{n=1}^{\infty} d_n \phi_{a_n}(x)$$

is increasing, the set of points at which f is discontinuous [denoted Disc(f)] is S, and $f(a_n+) - f(a_n-) = d_n$. [**Note:**

$$f(x) = \sum_{n=1}^{N} d_n \phi_{a_n}(x) + \sum_{n=N+1}^{\infty} d_n \phi_{a_n}(x)$$

and for every $\varepsilon > 0$, there exists N such that

$$\left| \sum_{n=N+1}^{\infty} d_n \phi_{a_n}(x) \right| < \varepsilon. \Bigg]$$

Show that the set of points where f is not differentiable is equal to Disc(f) = $\{a_n\}$.

9. **Definition:** Let f be an increasing function defined on the real line. We say that I is an *interval of constancy* for f if

(a) I contains more than one point,
(b) f is constant on I,
(c) there is no larger interval than I on which f is constant.

Prove that the set of intervals of constancy of an increasing function can be enumerated.

10. Let ϕ_0 be defined for all $x \in R^1$ such that ϕ_0 has period 1 and

$$\phi_0(x) = \begin{cases} x, & 0 \le x \le \dfrac{1}{2}, \\ 1 - x, & \dfrac{1}{2} \le x \le 1. \end{cases}$$

Let

$$\phi_n(x) = \frac{1}{4^n}\phi_0(4^n x).$$

The function ϕ_n has period 4^{-n} and it has a derivative a.e. equal to 1 or -1. Those points where the derivative does not exist are called *angular points* and have abscissae $\dfrac{p}{2^n}$. Let

$$f(x) = \sum_{n=1}^{\infty} \phi_n(x).$$

Show that f is continuous but lacks a derivative at every point. (Hint: Let x_0 and m be fixed and $h = \pm\dfrac{1}{4^m}$. Then $\phi_n(x_0 + h) - \phi_n(x_0) = 0$, $n \ge m$. The function ϕ_{m-1} has intervals without angular points of length $\dfrac{2}{4^m}$. The interval without angular points which contains x_0 will also contain one of the intervals $\left(x_0, x_0 + \dfrac{1}{4^m}\right)$ or $\left(x_0, x_0 - \dfrac{1}{4^m}\right)$. But all of the preceding functions ϕ_n, $n < m - 1$, have no angular points in this interval, so that

$$\frac{\phi_n(x_0 + h) - \phi_n(x_0)}{h} = 1, \quad n < m - 1.$$

Hence

$$\frac{f(x_0 + h) - f(x_0)}{h} = \sum_{n=0}^{m-1} \frac{\phi_n(x_0 + h) - \phi_n(x_0)}{h},$$

which is even if m is odd and odd if m is odd. Thus

$$\lim_{h \to 0} \frac{f(x_0 + h) - f(x_0)}{h}$$

does not exist.)

CHAPTER 2

Density and Approximate Continuity

The concepts of density and approximate continuity play a critical role in analysis of functions of several variables, with the important result being that for any set $E \subset \mathbf{R}^n$,

$$\lim_{r \to 0} \frac{m^*(B(x,r) \cap E)}{mB(x,r)} = 1$$

for almost all $x \in E$, m and m^* now denoting Lebesgue measure and Lebesgue outer measure, respectively, in $\mathbf{R}^n$. The result is also true if we replace Lebesgue measure by Hausdorff measure (see Chapter 8). In this chapter, we prove the above result (Theorem 2.2.1) in the setting of $\mathbf{R}^1$. We also show the connection between approximate continuity and integrability. Sierpinski's theorem extends the concept of approximate continuity to nonmeasurable functions. We end this chapter with a discussion of the Darboux property and the density topology.

§2.1. Preliminaries and Definitions

Definition 2.1.1. Let E be any set and I be an interval. The *relative measure of E in I* is given by the number

$$\frac{m^*(E \cap I)}{mI}.$$

Definition 2.1.2. For any real number x, let

$$\bar{\phi}(x) = \lim_{n \to \infty} \sup \left(\frac{m^*(E \cap I)}{mI} : x \in I, mI < \frac{1}{n} \right)$$

and

$$\underline{\phi}(x) = \lim_{n \to \infty} \inf \left(\frac{m^*(E \cap I)}{mI} : x \in I, mI < \frac{1}{n} \right).$$

The numbers $\bar{\phi}(x)$ and $\underline{\phi}(x)$ are called the *upper outer* (or *exterior*) *density* and the *lower outer* (or *exterior*) *density of E at x*, respectively. We observe that the value of the upper and lower densities would be unchanged if the intervals I were required to be open, closed, or half-closed. We will usually consider closed intervals in order to apply the Vitali covering theorem.

Definition 2.1.3 [S1, p. 117]. Let E be any set. Let

$$L_E(A) = m^*(E \cap A)$$

for every measurable set A. The function L_E is called the *measure function* for the set E.

We observe that L_E is an additive set function, and if E is a measurable set,

$$L_E(A) = \int_A \chi_E(x)\, dx$$

for every measurable set A; i.e., L_E is the indefinite integral of the characteristic function of the set E.

Definition 2.1.4 [M4, p. 287]. Let ψ be any function of a set. The *strong upper derivate of ψ at the point x* is the upper limit of the ratio

$$\frac{\psi(I_n)}{m(I_n)},$$

where I_n is any sequence of intervals, each of which contains the point x and whose diameters tend to 0. The *strong lower derivate of ψ at the point x* is similarly defined to be the lower limit of such ratios.

Thus, the strong upper derivate of the measure function of E is the upper outer density of E, whereas the strong lower derivate is the lower outer density of E.

Definition 2.1.5. If the upper and lower outer densities of a set E are equal at the point x, then their common value, $\phi(x) = \bar{\phi}(x) = \underline{\phi}(x)$, is called the *outer density of E at the point x*.

If the set E is measurable, then we replace outer measure m^* by measure m in the preceding definitions and replace outer density by density, or *metric* *density*, as some authors prefer. We shall simply use the term density, unless the context requires that we emphasize the measurability or nonmeasurability of the set in question.

The reader should verify that, for every set E and every real number x:

(2.1.6) $0 \leq \underline{\phi}(x) \leq \overline{\phi}(x) \leq 1$.

(2.1.7)   If $\overline{\phi}(x) = 0$, then the density of E at x exists and equals 0. If $\underline{\phi}(x) = 1$, then the density of E at x exists and is equal to 1.

EXAMPLE 2.1.8 [G6, p. 173]. This example illustrates the case where the upper and lower densities are not equal. Let E be the closed interval $[\frac{1}{2}, 1]$. For every positive integer n, let

$$I_n = \left(\frac{1}{2} - \frac{1}{n^2}, \frac{1}{2} + \frac{1}{n} \right).$$

Then

$$\frac{m(E \cap I_n)}{mI_n} = \frac{n}{n+1}.$$

For

$$I_n = \left(\frac{1}{2} - \frac{1}{n}, \frac{1}{2} + \frac{1}{n^2} \right),$$

$$\frac{m(E \cap I_n)}{mI_n} = \frac{1}{n+1}.$$

Thus, the upper density of E at $\frac{1}{2}$ is 1 and the lower density of E at $\frac{1}{2}$ is 0; hence, the density of E does not exist at $\frac{1}{2}$.

EXAMPLE 2.1.9 [G6, p. 173]. We relate another example of interest.

$$E = \bigcup_{n=1}^{\infty} \left(\frac{1}{2} + \frac{1}{n}, \frac{1}{2} + \frac{1}{n} + \frac{1}{n^2} \right).$$

Then

$$\frac{m(E \cap I)}{mI} \geq 0$$

for every open interval I containing $\frac{1}{2}$, but the density of E at $\frac{1}{2}$ exists and is 0.

Definition 2.1.10. If the density of E at the point x is 1, then x is called a *point of density* of the set E. If the density of E at the point x is 0, then x is called a *point of dispersion* (or *point of rarefaction*) of E.

Theorem 2.1.11 [G6, p. 173]. *If E is measurable and the density of E exists at x, then the density of E^C exists at x and the sum of the two densities is equal to 1.*

Proof. For every interval I,

$$m(E \cap I) + m(E^C \cap I) = mI,$$

so that

$$\frac{m(E \cap I)}{mI} + \frac{m(E^C \cap I)}{mI} = 1.$$

The result now follows from the definition. $\qquad\qquad\qquad\qquad\square$

Corollary 2.1.12. *If the set E is measurable, then any point of density of E is a point of dispersion for E^C.*

§2.2. The Lebesgue Density Theorem

Theorem 2.2.1 (Lebesgue Density Theorem [J1, p. 114]). *Almost all points of an arbitrary set E are points of density for E.*

Proof. Suppose that the theorem is not true. Then there exists a set $B \subset E$ with $m^*(B) > 0$, and such that at each $x \in B$, $\phi(x) < \lambda < 1$. If $x \in B$, there is associated with x a sequence of closed intervals I_n, such that $x \in I_n$ for each n,

$$\lim_{n \to \infty} mI_n = 0$$

and

$$\frac{m^*(E \cap I_n)}{mI_n} < \lambda.$$

Since $B \subset E$, then for $x \in B$

$$\frac{m^*(B \cap I_n)}{mI_n} < \lambda.$$

These intervals I_n associated with the points $x \in B$ form a Vitali covering of the set B. Hence, for an arbitrary $\varepsilon > 0$, there exists a finite set, $\{I_1, I_2, \ldots, I_j\}$, of these intervals which are mutually disjoint such that

$$m^* \left[\sum_{n=1}^{j} (B \cap I_n) \right] > m^*(B) - \varepsilon$$

and

$$\sum_{n=1}^{j} mI_n < m^*(B) + \varepsilon.$$

Since the intervals $I_1, \ldots, I_j$ are mutually disjoint, we have

$$m^* \left(\sum_{n=1}^{j} (B \cap I_n) \right) = \sum_{n=1}^{j} m^*(B \cap I_n).$$

Combining these results, we get

$$m^*(B) - \varepsilon < \sum_{n=1}^{j} m^*(B \cap I_n) < \lambda \sum_{n=1}^{j} mI_n < \lambda[m^*(B) + \varepsilon].$$

Since $\lambda < 1$ and ε can be made arbitrarily small, we get a contradiction. Thus, $m^*(B) = 0$. If $\lambda_1 < \lambda_2 < \cdots$ is a sequence of numbers such that $\lambda_n \to 1$, then the set $B_n \subset E$ at which the lower density of E is less than λ_n is a set of measure zero. But the points in E at which the lower density is less than 1 is the set

$$B = B_1 \cup B_2 \cup \cdots.$$

Hence,

$$m^*(B) \leq m^*(B_1) + m^*(B_2) + \cdots = 0.$$

Thus, $m^*(B) = 0$; therefore, $mB = 0$. □

Corollary 2.2.2 [G6, p. 175]. *If E is a measurable set, the density of E exists and is equal to 0 or 1 at every point, except for a set of measure zero.*

Proof. By Theorem 2.2.1, $E = B_1 \cup Z_1$ where the density of E exists and is equal to 1 at every point of B_1 and $mZ_1 = 0$. Also, $E^C = B_2 \cup Z_2$, where the density of E^C exists and is equal to 1 at every point of B_2 and $mZ_2 = 0$. By Theorem 2.1.11, the density of E exists and is 0 at every point of B_2. Hence, the density of E exists and is 0 or 1 everywhere except at points of $Z = Z_1 \cup Z_2$, which is a set of measure zero. □

We present now the converse of the Lebesgue density theorem.

Theorem 2.2.3 [G6, pp. 175–176]. *If Z is a set of measure zero, there is a measurable set E such that the density of E does not exist at any point of Z.*

Proof. Let $G_1 \supset G_2 \supset \cdots G_n \supset \cdots$ be a sequence of open sets each covering Z. Each G_n being an open set, it may be written as the union of a countable set of disjoint open intervals:

$$G_n = \bigcup_{p=1}^{\infty} I_{np}.$$

Since $mZ = 0$, the sets G_n may be chosen such that the relative measure of G_{n+1} is $1/n$ in each of the disjoint intervals I_{np} whose union is G_n; i.e.,

$$\frac{m(G_{n+1} \cap I_{np})}{mI_{np}} = \frac{1}{n}.$$

Let

$$E = (G_1 - G_2) \cup (G_3 - G_4) \cup \cdots \cup (G_{2k-1} - G_{2k}) \cup \cdots.$$

Then E is measurable. Let $z \in Z$. For every n, there is a p_n such that $z \in I_{np_n}$, one of the disjoint intervals in G_n. For odd values of n, $G_n - G_{n+1} \subset E$. Hence,

$$\frac{m(I_{np_n} \cap (G_n - G_{n+1}))}{mI_{np_n}} \leq \frac{m(I_{np_n} \cap E)}{mI_{np_n}}.$$

But

$$I_{np_n} \cap (G_n - G_{n+1}) = I_{np_n} \cap (G_{n+1})^C.$$

Since

$$m(G_{n+1} \cap I_{np_n}) + m((G_{n+1})^C \cap I_{np_n}) = mI_{np_n},$$

we have

$$\frac{m(I_{np_n} \cap E)}{mI_{np_n}} \geq 1 - \frac{1}{n}.$$

If n is even, $E \cap I_{np_n} \subset G_{n+1}$. Hence,

$$\frac{m(E \cap I_{np_n})}{mI_{np_n}} \leq \frac{m(I_{np_n} \cap G_{n+1})}{mI_{np_n}} = \frac{1}{n}.$$

But the length of I_{np_n} converges to 0 as n increases for both even and odd values of n. Thus, the upper density of E at z is 1 and the lower density of E at z is 0, so that the density of E does not exist at any point $z \in Z$. $\qquad\square$

Finally, we present a category result for the set of points where the density of a given set is neither 0 nor 1.

Theorem 2.2.4 [G6, p. 176]. *The set of points at which the density of a measurable set E exists but is not equal to 0 or 1 is of measure zero and of the first category.*

Proof. Let T be the set of points for which the density of E exists. Let U be those points of T for which the density of E is either 0 or 1. Let $Z = T - U$. By Theorem 2.2.1, $mZ = 0$. For every positive integer n and for every x, define the function

$$f_n(x) = \frac{m(E \cap (x - 1/n, x + 1/n))}{2/n}.$$

It is easy to see that f_n is a continuous function. For every $x \in T$, the density of E exists, so

$$f(x) = \lim_{n \to \infty} f_n(x)$$

exists for every $x \in T$ and is equal to the density of E at x. Thus, f is of Baire class 1 on T relative to T and its points of discontinuity must be a set of first category relative to T. Also, U is dense in T. Thus, every interval containing a point of Z also contains a point of U at which f is either 0 or 1. Since, at every point of Z, f is neither 0 or 1, Z is a subset of the points of discontinuity of f. Hence, Z is of first category. □

Now we give a result which characterizes the measurability of a set in terms of density.

Theorem 2.2.5 [M4, p. 290]. *A necessary and sufficient condition that a set E be measurable is that x be a point of dispersion of E for almost all $x \in E^C$.*

Proof. If E is measurable, then almost all points of E^C are points of density for E^C, hence points of dispersion for E (by Corollary 2.1.12). □

If E is not measurable, then there exists a measurable set M such that $E \subset M$ and $m^*(E) = mM$. Almost all points of $M - E$ are points of density of M and, hence, for E because, for any interval I,

$$m(M \cap I) = mM - M(M - I)$$
$$= m^*(E) - m(M - I)$$
$$\leq m^*(E) - m^*(E - I)$$
$$\leq m^*(E \cap I).$$

Since $E \subset M$, we have $m^*(E \cap I) \leq m(M \cap I)$; hence, $m^*(E \cap I) = m(M \cap I)$. But $m^*(M - E) > 0$; thus, the set of points of E^C which are not points of dispersion for E is a set with positive outer measure.

§2.3. Approximate Continuity

The concept of approximate continuity was first introduced by Denjoy in 1915. We begin this section by stating several equivalent definitions.

Definition 2.3.1. A function f is said to be *approximately continuous* at a point x_0 if, for every $\varepsilon > 0$, the set

$$\{x: |f(x) - f(x_0)| < \varepsilon\}$$

has density 1 at the point x_0.

Theorem 2.3.2 [H2, pp. 235–236]. *A necessary and sufficient condition that f should be approximately continuous at the point x_0 is that there exists a set E such that x_0 is a point of density of E and $f|_E$ is continuous at x_0.*

Proof. Suppose there exists a set E such that x_0 is a point of density of E and $f|_E$ is continuous at x_0. Then, for every $\varepsilon > 0$, there exists $h > 0$ such that if

$x \in (x_0 - h, x_0 + h) \cap E$, then $|f(x) - f(x_0)| < \varepsilon$. Since the density of the set E and the set $E \cap (x_0 - h, x_0 + h)$ at the point x_0 is the same, we have proved the sufficiency.

Now suppose $\varepsilon > 0$ and let $E(x_0, \varepsilon)$ be the set of points at which $|f(x) - f(x_0)| < \varepsilon$ has density 1 at the point x_0. Let $\{\varepsilon_n\}$ be a sequence of decreasing values of ε such that $\varepsilon_n \to 0$. For each ε_n, we can find a number h_n $(<h_{n-1})$ such that

$$m[E(x_0, \varepsilon_n) \cap (x_0 - h_n, x_0 + h_n)] > 2h_n(1 - \varepsilon_n).$$

We can also choose the numbers h_n such that

$$\frac{h_{n+1}}{h_n} < \frac{1}{\varepsilon_n}.$$

Let F_n be the part of $E(x_0, \varepsilon_n)$ that is contained in the two intervals $(x_0 + \varepsilon_n h_{n+1}, x_0 + h_n)$ and $(x_0 - h_n, x_0 - \varepsilon_n h_{n+1})$. Then $F_{n+m} \subset E(x_0, \varepsilon_{n+m}) \subset E(x_0, \varepsilon_n)$. Let

$$E = \bigcup_{n=1}^{\infty} F_n.$$

Then E has density 1 at the point x_0, since $mF_n > 2h_n(1 - 2\varepsilon_n)$. Let $x_1, x_2, \ldots,$ $x_n, \ldots$ be a sequence of numbers (increasing or decreasing) which converge to x_0, such that each $x_m \in E$. Each x_m belongs to one of the sets $F_1, F_2, \ldots, F_n,$ $\ldots,$ say $F_{\bar{m}}$. Then $\bar{m} \to \infty$ as $m \to \infty$. Thus, we have $|f(x_m) - f(x_0)| < \varepsilon_{\bar{m}}$; therefore, the sequence $\{f(x_m)\}$ converges to $f(x_0)$. Hence, f is continuous relative to E and E has density 1 at x_0. $\square$

Remark. Theorem 2.3.2 is also used as the definition of approximate continuity [B2, p. 18]. It will be left as an exercise for the reader to confirm that the following definitions of approximate continuity are equivalent to Definition 2.3.1.

Definition 2.3.3 [J1, p. 118]. Let f be defined on a set E and let x_0 be a point of density of E. If

$$\lim_{x \to x_0} f(x) = f(x_0)$$

except for a set with density 0 at x_0, then f is said to be approximately continuous at x_0.

Definition 2.3.4 [S1, pp. 131–132]. The function f is approximately continuous at x_0 if $f(x_0) \neq \infty$ and $f(x) \to f(x_0)$ as $x \to x_0$ on a measurable set E for which x_0 is a point of density.

Definition 2.3.5 [M4, p. 291]. The function f is approximately continuous at x_0 if, for every $\varepsilon > 0$, x_0 is a point of dispersion for the set

$$\{x : |f(x) - f(x_0)| \geq \varepsilon\}.$$

Definition 2.3.6 [G6, p. 189]. A measurable function f is said to be approximately continuous at x_0 if for every pair of real numbers k_1 and k_2 such that $k_1 < f(x_0) < k_2$, the set

$$\{x: k_1 < f(x) < k_2\}$$

has density 1 at x_0.

Definition 2.3.7 [G6, p. 189]. The function f is approximately continuous at x_0 if for every pair of rational numbers r_1 and r_2 such that $r_1 < f(x_0) < r_2$ the set

$$\{x: r_1 < f(x) < r_2\}$$

has density 1 at x_0.

Theorem 2.3.8. *If a function f is continuous at x_0, then f is approximately continuous at x_0.*

Proof. If f is continuous at x_0, then for every $\varepsilon > 0$, the set

$$\{x: |f(x) - f(x_0)| < \varepsilon\}$$

contains a neighborhood of x_0 and thus has density 1 at x_0.   □

That the converse of Theorem 2.3.8 is not true is illustrated by the following examples of functions which are approximately continuous but not continuous.

EXAMPLE 2.3.9 [M4, p. 292]. Let f be the characteristic function of the rationals. This function is approximately continuous at every irrational number but is discontinuous everywhere.

EXAMPLE 2.3.10 [G6, p. 190]. Let f be defined on $[-1, 1]$ as follows: From each interval

$$\left(\frac{1}{n+1}, \frac{1}{n}\right), \quad n = 1, 2, \ldots$$

remove a subinterval of length

$$\frac{1}{n}\left(\frac{1}{n} - \frac{1}{n+1}\right) = \frac{1}{n^2(n+1)}.$$

Let E be the union of all the intervals removed and let

$$f(x) = \begin{cases} 1, & x \in E \\ 0, & x \in E^C. \end{cases}$$

Define f on $[-1, 0]$ by symmetry. This function is approximately continuous but not continuous at 0.

Theorem 2.3.11 [B2, p. 20]. *If f and g are approximately continuous at x_0, then $f + g, f \cdot g, f - g$, and f/g (if $g \neq 0$) are approximately continuous at x_0.*

Proof. The proof is an easy consequence of the following: If the density of the set E_f at x_0 is 1 and the density of the set E_g at the point x_0 is 1, the density of the set $E_f \cap E_g$ at x_0 is 1. □

Theorem 2.3.12 [B2, p. 20]. *If f is approximately continuous at x_0 and g is continuous at $f(x_0)$, then $g \circ f$ is approximately continuous at x_0.*

Proof. This result follows from the definition of approximate continuity given by Theorem 2.3.2. □

The order of composition in Theorem 2.3.12 is important; it is possible for f to be approximately continuous and g continuous, whereas $f \circ g$ is not approximately continuous [B2, p. 20].

Theorem 2.3.13 [M4, p. 292]. *A necessary and sufficient condition that f be measurable is that it be approximately continuous a.e.*

Proof. Suppose f is measurable. Then by Lusin's theorem, given $\varepsilon > 0$, there is a closed set F such that $mF^C < \varepsilon$, and $f|_F$ is continuous. If x is a point of dispersion for F^C, x is a point of density of F and f is approximately continuous at x. Hence, f is approximately continuous a.e. in F. So the set of points at which f is not approximately continuous has measure less than ε. Since ε is arbitrary, this set must have measure zero. Thus, we have proved the necessity.

Now suppose f is approximately continuous a.e. Let a be any real number and let

$$E = \{x: f(x) > a\}.$$

If $x_0 \in E$ and is a point of approximate continuity for f, let $\varepsilon = f(x_0) - a > 0$. Then x_0 is a point of dispersion for the set

$$S = \{x: |f(x) - f(x_0)| \geq \varepsilon\}.$$

Thus, x_0 is a point of dispersion for E^C since $E^C \subset S$. Since this is true for almost all $x_0 \in E$, then E^C is measurable (by Theorem 2.2.5), and hence E is measurable. Therefore, f is measurable. □

§2.4. Approximate Continuity and Integrability

For a bounded function on a bounded interval, we know that a necessary and sufficient condition for the function to be Riemann integrable is that it be continuous a.e. Theorem 2.3.13 tells us that a bounded function on a bounded interval is Lebesgue integrable if and only if it is approximately continuous a.e.

Theorem 2.4.1 [B2, p. 21]. *Let f be a bounded and approximately continuous function defined on an interval I. Let $a \in I$ and define*

$$F(x) = \int_a^x f(t)\,dt,$$

then $f = F'$.

Proof. We will show that for each $x_0 \in I$,

$$\lim_{h \to 0} \frac{1}{h} \int_{x_0}^{x_0+h} f(t)\,dt = f(x_0).$$

Let $x_0 \in I$ and let $E \subset I$ be such that the density of E at x_0 is 1 and $f|_E$ is continuous at x_0. Let M be an upper bound for $|f|$. For every $h > 0$,

$$\left| \frac{1}{h} \int_{x_0}^{x_0+h} f(t)\,dt - f(x_0) \right| = \left| \frac{1}{h} \int_{x_0}^{x_0+h} [f(t) - f(x_0)]\,dt \right|$$

$$\leq \frac{1}{h} \int_{x_0}^{x_0+h} |f(t) - f(x_0)|\,dt$$

$$= \frac{1}{h} \int_{[x_0,x_0+h] \cap E} |f(t) - f(x_0)|\,dt$$

$$+ \frac{1}{h} \int_{[x_0,x_0+h] - E} |f(t) - f(x_0)|\,dt.$$

For $\varepsilon > 0$, choose $\delta > 0$ such that if $t \in E$ and $|t - x_0| < \delta$, then $|f(t) - f(x_0)| < \varepsilon/2$ and if $h < \delta$, then

$$\frac{m([x_0, x_0 + h] - E)}{h} < \frac{\varepsilon}{4M}.$$

Then, for $h < \delta$, we have

$$\left| \frac{1}{h} \int_{x_0}^{x_0+h} f(t)\,dt - f(x_0) \right| \leq \frac{\varepsilon}{2h} m([x_0, x_0 + h] \cap E) + \frac{2M}{h} m([x_0, x_0 + h] - E)$$

$$\leq \frac{\varepsilon}{2h} h + \frac{2M}{h} \frac{\varepsilon h}{4M} = \varepsilon.$$

A similar calculation holds for $h < 0$. Thus,

$$\lim_{h \to 0} \frac{1}{h} \int_{x_0}^{x_0+h} f(t)\,dt = f(x_0),$$

i.e., $F'(x_0) = f(x_0)$. □

Corollary 2.4.2 [M4, p. 293]. *If f is bounded and integrable, and F is its indefinite integral, then $F' = f$ a.e.*

If f is integrable but unbounded, approximate continuity is not sufficient to ensure that $F' = f$ (see Exercise 7 at the end of this chapter).

§2.5. Further Results on Approximate Continuity

Theorem 2.5.1 [B2, p. 23]. *Let f be defined on an interval [a, b]. For each $\alpha \in \mathbf{R}^1$, let*

$$E^\alpha = \{x: f(x) < \alpha\},$$

$$E_\alpha = \{x: f(x) > \alpha\}.$$

If each point of E^α is a point of density of E^α and each point of E_α is a point of density of E_α, then f is approximately continuous.

Proof. Let $x_0 \in [a, b]$. For every $\varepsilon > 0$, x_0 is a point of density of the sets

$$\{x: f(x) < f(x_0) + \varepsilon\},$$

$$\{x: f(x) > f(x_0) - \varepsilon\}$$

and therefore of their union

$$\{x: |f(x) - f(x_0)| < \varepsilon\}.$$

Hence, f is approximately continuous at each $x_0 \in [a, b]$. $\qquad \square$

In [B2, p. 23] it is shown that Theorem 2.5.1 is a necessary as well as a sufficient condition for approximate continuity.

Theorem 2.5.2 [B2, p. 24]. *Let $\{f_n\}$ be a sequence of approximately continuous functions defined on the interval $I = [a, b]$. If $f_n \to f$ uniformly, then f is approximately continuous.*

Proof. Let $\alpha \in \mathbf{R}^1$. We will show that each point of

$$E^\alpha = \{x: f(x) < \alpha\}$$

is a point of density of E^α. Let $x_0 \in E^\alpha$. Choose N such that

$$|f_N(x) - f(x)| < \frac{\alpha - f(x_0)}{3}$$

for all $x \in I$. Let

$$E_N = \left\{|f_N(x) - f_N(x_0)| < \frac{\alpha - f(x_0)}{3}\right\}.$$

Then the density of E_N at x_0 is 1 because f_N is approximately continuous. Since $E_N \subset E^\alpha$, x_0 is a point of density of E^α. A similar argument shows that each point of E_α is a point of density of E_α. Thus, by the previous theorem, f is approximately continuous. □

§2.6. Sierpinski's Theorem

Definition 2.6.1 [S3]. Let f be a function (not necessarily measurable) defined over a measurable set E. We will say that f satisfies *property P* at x_0 if, for any $\varepsilon > 0$, the set

$$E(x_0, \varepsilon) = \{x : |f(x) - f(x_0)| < \varepsilon\}$$

has outer density 1 at x_0.

Remark. Certainly if f is measurable, the sets $E(x_0, \varepsilon)$ are also measurable and the points of outer density 1 coincide with points of density. Hence, for measurable functions, property P coincides with approximate continuity. Thus, property P may be seen as an extension of the concept of approximate continuity to arbitrary functions, measurable or not.

Theorem 2.6.2 [S3]. *Every function f (measurable or not) satisfies property P a.e.*

Proof. Let

$$E_k^n = \left\{x : \frac{k}{n} \le f(x) < \frac{k+1}{n}\right\}.$$

For every positive integer n,

$$\mathbf{R}^1 = E_0^n \cup E_1^n \cup E_{-1}^n \cup E_2^n \cup E_{-2}^n \cup \cdots.$$

By the Lebesgue density theorem (Theorem 2.2.1), almost all points of E_k^n are points of density for E_k^n. Let

$$E_k^n = H_k^n \cup N_k^n,$$

where $m(N_k^n) = 0$ and the points of H_k^n are points of density of E_k^n. Then

$$\mathbf{R}^1 = N_0^n \cup N_1^n \cup N_{-1}^n \cup \cdots \cup H_0^n \cup H_1^n \cup H_{-1}^n \cup \cdots.$$

Let

$$N^n = N_0^n \cup N_1^n \cup N_{-1}^n \cup N_2^n \cup N_{-2}^n \cup \cdots.$$

Then $mN^n = 0$. Now let $x_0 \in \mathbf{R}^1 - (\cup N^n)$ and let $\varepsilon > 0$. Choose n such that $n > 1/\varepsilon$. Since $x_0 \in \mathbf{R}^1 - (\cup N^n)$, $x_0 \in \mathbf{R}^1 - N^n$. Then there exists an integer k

such that $x_0 \in H_k^n$ and thus x_0 is a point of density of E_k^n. Recall

$$E_k^n = \left\{ x : \frac{k}{n} \leq f(x) < \frac{k+1}{n} \right\}.$$

Thus, for $x \in E_k^n$,

$$|f(x) - f(x_0)| < \frac{1}{n} < \varepsilon.$$

This implies that the set

$$E(x_0, \varepsilon) = \{ x : |f(x) - f(x_0)| < \varepsilon \} \supset E_k^n$$

has x_0 as a point of outer density. Since ε is arbitrary, f has the property P at x_0, and hence a.e., as $\cup N^n$ has measure 0. $\qquad\square$

§2.7. The Darboux Property and the Density Topology

Definition 2.7.1. A real-valued function f defined on an interval I satisfies the *intermediate value property* (or *Darboux property*) if whenever $x_1, x_2 \in I$, and y is any number between $f(x_1)$ and $f(x_2)$, then there exists a number x_3 between x_1 and x_2 such that $y = f(x_3)$. Functions having the intermediate value property are called *Darboux functions* because of Darboux's work on the subject. He showed that the intermediate value property is not equivalent to continuity and that every derivative has the intermediate value property (see exercise 14).

EXAMPLE 2.7.2 [B2, p. 3]. Let C be the Cantor ternary set. If (a, b) is an interval contiguous to C, define

$$f(x) = \begin{cases} \dfrac{2(x-a)}{b-a} - 1, & x \in [a, b] \\[2mm] 0, & \text{otherwise.} \end{cases}$$

Then f is a Darboux function which is discontinuous at every point of C.

In [GW], an equivalent statement of Definition 2.7.1 is given: If $\alpha, \beta \in I$, $\alpha < \beta$, the image $f((\alpha, \beta))$ of (α, β) contains the open interval whose endpoints are $f(\alpha)$ and $f(\beta)$. Another equivalent definition of the Darboux property is given in the following theorem.

Theorem 2.7.3 [GW]. *A real-valued function f defined on an interval I has the Darboux property if and only if f takes every connected set into a connected set.*

Proof. If f takes every connected set into a connected set, then clearly f has the Darboux property. Conversely, suppose f has the Darboux property and let S be a connected subset of I. Then S is an interval. If $f(S)$ is not connected, i.e., $f(S)$ is not an interval, then there exist $c < \eta < d$ with $c, d \in f(S)$ and $\eta \notin f(S)$. But then the image of the open interval with endpoints $f^{-1}(c)$ and $f^{-1}(d)$ does not contain (c, d). This contradiction implies $f(S)$ must be connected.   □

Another condition equivalent to the Darboux property given in [GW] is that f takes closed intervals into connected sets. The following example shows the condition that f takes open intervals into connected sets is not equivalent to the Darboux property.

EXAMPLE 2.7.4 [GW]. Let f be defined on $(-1, 1)$:

$$f(x) = \begin{cases} 0, & x < 0 \\ 1, & x = 0 \\ \sin\left(\dfrac{1}{x}\right), & x > 0. \end{cases}$$

Clearly, f takes open intervals into connected sets, but f does not have the Darboux property.

Theorem 2.7.5 [B2, p. 22]. *Every approximately continuous function has the Darboux property.*

Proof. If f is bounded and F is defined as in Theorem 2.4.1, then $f = F'$. Thus, f has the Darboux property. If f is not bounded, let h be a homeomorphism of $\mathbf{R}^1$ onto $(0, 1)$. Then $h \circ f$ is bounded and approximately continuous (Theorem 2.3.12); hence, $h \circ g$ has the Darboux property and thus $f = h^{-1} \circ h \circ f$ has the Darboux property.   □

Definition 2.7.6 [GW]. A measurable set is called *homogeneous* if its density is 1 at each of its points.

We note that the union or finite intersection of homogeneous sets is homogeneous and that the space $\mathbf{R}^1$ and the empty set are homogeneous. Thus, $\mathbf{R}^1$ can be topologized by taking the homogeneous sets as open sets. This topology on $\mathbf{R}^1$, called the *density topology or the d-topology*, is Hausdorff.

In [GW], we find yet another equivalent definition of approximate continuity: f is approximately continuous at x_0 if, for every open set G containing $f(x_0)$, the set $f^{-1}(G)$ has density 1 at x_0. Thus, if $f: \mathbf{R}^1 \to \mathbf{R}^1$ is approximately continuous, then with the usual topology on the range and the d-topology on the domain, f is continuous.

Definition 2.7.7 [GW]. A point p is called a *d-limit point* of the set E, i.e., p is a limit point of the set E in the d-topology, if E has positive upper density at p. A *d-connected set*, i.e., a set which is connected in the d-topology, is a set which is not the union of two nonempty subsets neither of which contains a d-limit point of the other.

Theorem 2.7.8 [GW]. *Every open connected subset of $\mathbf{R}^1$ is d-connected.*

Proof. Suppose E is an open connected set and there exist sets A and B, nonempty, disjoint, homogeneous, and such that $E = A \cup B$. Let $p_0 \in A$, $p_1 \in B$, and C be a simple arc in E connecting p_0 and p_1. For every point p, let $I(p, r)$ be the closed interval of length r with midpoint p. For any measurable set, the relative measure of the set $I(p, r)$ is a continuous function of p for fixed r.

Choose r so that $I(r, p) \subset E$ for every $p \in C$ and so that the relative measure of A in $I(p, r)$ exceeds $\frac{1}{2}$ for $p = p_0$ and is less than $\frac{1}{2}$ for $p = p_1$. This is possible since A has density 1 at p_0 and density 0 at p_1. There must then be a point $\bar{p} \in C$ such that the relative measure of A in $I(\bar{p}, r)$ is exactly $\frac{1}{2}$. Let $I_1 = I(\bar{p}, r)$. I_1 must have interior points $p_0^1 \in A$ and $p_0^1 \in B$. Join these by a line segment C^1. We may choose $r^1 < r/2$ so that $I(p, r_1) \subset I_1$ for every $p \in C^1$ and so that the relative measure of A in $I(p, r^1)$ exceeds $\frac{1}{2}$ for $p = p_0^1$ and is less than $\frac{1}{2}$ for $p = p_1^1$. There is then a point $p^* \in C^1$ such that the relative measure of A in $I(p^*, r^1)$ is exactly $\frac{1}{2}$. Let $I_2 = I(p^*, r^1)$. Continuing this process, we construct a nested sequence $\{I_n\}$ of closed intervals in each of which the relative measure of A is $\frac{1}{2}$. This sequence converges to a point $p' \in E$, but neither A nor B can have density 1 at p'. Since A and B are homogeneous, $p' \notin A$ and $p' \notin B$. This contradiction proves the result. $\qquad\square$

Definition 2.7.9 [GW]. A closed set with connected interior whose boundary points are d-limit points of the interior is called a *d-regular set*. Closed convex sets are examples of d-regular sets.

Theorem 2.7.10 [GW]. *An approximately continuous function from $\mathbf{R}^1$ to a metric space takes d-regular sets into connected sets.*

Proof. Suppose that a d-regular set S is not d-connected. Then $S = A \cup B$ such that A and B are disjoint, nonempty, and each contains no d-limit point of the other. One of the sets, say A, must then be contained entirely in the boundary of S, for otherwise this separation would induce a separation in the interior of S, contradicting Theorem 2.7.8. Then A contains a d-limit point of the interior of S and hence of B, which is a contradiction. $\qquad\square$

EXERCISES

1. If E is a measurable set and k is a real number, the set of points at which the lower density of E is less than k is measurable.

2. Let E be the set in $\mathbf{R}^1$ made up of the intervals

$$\left[\frac{1}{2^{2k+1}}, \frac{1}{2^{2k}}\right], \quad k = 1, 2, 3, \ldots$$

and their reflections through the origin. Show that at 0, E has upper density $\frac{2}{3}$ and lower density $\frac{1}{3}$.

3. Show that if x is a point of dispersion for S, then x is a point of density for S^C. (Compare to Theorem 2.1.11 and Corollary 2.1.12.)

4. Show that the converse of Exercise 3 holds only if S is measurable.

5. Let E be a set with density $\frac{1}{2}$ at x and let $f = \chi_E - \chi_{E^C}$. Show that

$$D\left(\int f(x)\right) = 0$$

but f is not approximately continuous at x.

6. Let $E \subset \mathbf{R}^1$ and suppose there exists $\alpha \in (0, 1)$ such that for every interval (a, b) we have $E \cap (a, b)$ can be covered by countably many intervals, the sum of whose lengths is less than or equal to $\alpha(b - a)$. Show that $m(E) = 0$; that is, a set that covers at most a fixed fraction of every interval covers almost none of every interval.

   How does this result relate to density? The main result (Theorem 2.2.1) of Section 2.2 that "almost all points of a set are points of density of that set" can be interpreted as "a set just about fills up small neighborhoods of its points but cannot, for example, fill up about half of every interval."

7. Let f be defined by

$$f(x) = \begin{cases} 2^n, & \dfrac{1}{2^n} - \dfrac{1}{2^{2n+1}} \le x \le \dfrac{1}{2^n}; n = 0, 1, 2, \ldots \\ 0, & \text{otherwise.} \end{cases}$$

Let $\sigma = \int f$. Show f is approximately continuous at 0 but σ is not differentiable at 0. [**Hint:** Let $E = \{x: f(x) \ne 0\}$. If the interval J has diameter less than or equal to $1/2^n$ and $0 \in J$, show

$$\frac{m(E \cap J)}{mJ} \to 0$$

as $n \to \infty$, so f is approximately continuous at 0. However, if $J_n = [0, 1/2^n]$, then

$$\frac{\int_{J_n} f}{mJ_n} \ne f(0).]$$

8. Show the equivalence of the definitions of approximate continuity.

9. If $A \subset \mathbf{R}^1$ and $B \subset \mathbf{R}^1$, let $A + B = \{a + b: a \in A, b \in B\}$. Suppose $m(A) > 0$ and $m(B) > 0$. Prove that $A + B$ contains an interval. [**Hint:** Use the existence of points of density in A and B; or let

$$f(x) = \int_{-\infty}^{\infty} \chi_A(x - t)\chi_B(t)\, dt;$$

show that f is continuous and that $\displaystyle\int_{\mathbf{R}^1} f \ne 0.]$

10. Let C be the Cantor ternary set. Use the results of problem 9 to show that $C + C$ contains an interval even though $mC = 0$.

11. This problem provides another approach to the Lebesgue Density Theorem. We will suppose that E lies in some compact interval since only small neighborhoods of points of E are relevant. Let $f(x) = m^*(E \cap (-\infty, x))$. Then f is an increasing function. Show that $f'(x) = 1$ for almost all $x \in E$.

First consider a fixed covering of E by countable open intervals $\{I_n\}$. Let

$$g(x) = \sum_n m(I_n \cap (-\infty, x)).$$

If $x \in E$ and h is small, then $x + h$ will be in one of the open intervals covering E; thus, $g(x + h) - g(x) = h$. Therefore, $g'(x) = 1$ if $x \in E$.

Consider a sequence of coverings of E whose total lengths $\alpha_n \to mE$ and

$$\sum_n [\alpha_n - mE]$$

is convergent. Then the functions g_n (defined as g above for the nth covering) are such that $g_n \to f$ and $g_n(x) - f(x) \le \alpha_n - mE$. Furthermore, $g_n - f$ is increasing and $\sum [g_n(x) - f(x)]$ is convergent. Thus, by Fubini's theorem, $\sum [g'_n(x) - f'(x)]$ is convergent a.e. Hence, $g'_n(x) - f'(x) \to 0$ a.e. But since $g'_n(x) = 1$ for $x \in E$, we have $f'(x) = 1$ a.e. in E.

12. Let S be a compact subset of $\mathbf{R}^1$. For any $x \notin S$, $d(x, S) > 0$ and is attained at some point of S. For $r > 0$, let $S_r = \{x \notin S: d(x, S) = r\}$. Show that $mS_r = 0$.

13. A point ξ is said to be a point of *asymptotic continuity* of a function f, measurable over an interval $[a, b]$, if there exists a measurable set E on which f is continuous and for which ξ is a point of density. Prove that almost every point $\xi \in [a, b]$ is a point of asymptotic continuity of f. [**Hint:** Every density point of a set E_ε ($mE_\varepsilon < b - a - \varepsilon$) on which f is continuous satisfies the condition.]

14. Let $f: [a, b] \to \mathbf{R}^1$ be differentiable on $[a, b]$ (f' need not be continuous) and suppose that γ is a number strictly between $f'_+(a)$ and $f'_-(b)$. Then there exists $\xi \in (a, b)$ such that $f'(\xi) = \gamma$. [**Hint:** Letting $g(x) = f(x) - \gamma x$, we see that $g'_+(a)$ and $g'_-(b)$ have strictly opposite signs and so g is not monotone on $[a, b]$. If $a \le \alpha < \beta \le b$ and $g(\alpha) = g(\beta)$, then Rolle's Theorem provides ξ.)

15. To illustrate the previous exercise, let $P(x) = (x^2 - 1)^2$ and define f on $[0, 1]$ by $f(0) = 0$ and $f(x) = n^{-3/2} P(2n(n + 1)x - 2n - 1)$ if n is a positive integer and $(n + 1)^{-1} \le x \le n^{-1}$. Then f is differentiable on $[0, 1]$ and f' is not continuous or even bounded on $[0, 1]$. In fact, $f'_+(0) = 0$ and if $b_n = (4n + 1)/4n(n + 1)$, then $b_n \to 0$ and $f'(b_n) \to \infty$. Thus f' attains every positive value γ on every interval $[0, b_n]$.

CHAPTER 3

Dini Derivatives

In this chapter, we return to the problem of derivatives of functions of a single variable. We first introduce Dini derivatives and present many of the standard results on their properties. We then relate the interesting example of Ruziewicz showing that the difference of two functions with equal derivatives is not a constant if the derivative is not finite. We show the role Dini derivatives play in determining the monotonicity of a function. We completely characterize the Dini derivatives of any function in the Denjoy–Saks–Young Theorem. Banach's original proof of the measurability of the Dini derivatives of a measurable function is given. The chapter ends with some results on the Dini derivatives of a convex function.

§3.1. Preliminaries and Definitions

Definition 3.1.1. Let f be a real-valued function defined and finite on an interval $[a, b]$ and let $x_0 \in [a, b]$. Then

(3.1.2)   *the upper derivate of f at x_0 is*

$$\overline{D}f(x_0) = \limsup_{x \to x_0} \frac{f(x) - f(x_0)}{x - x_0};$$

(3.1.3)   *the lower derivate of f at x_0 is*

$$\underline{D}f(x_0) = \liminf_{x \to x_0} \frac{f(x) - f(x_0)}{x - x_0}.$$

If $x_0 < b$,

(3.1.4) *the upper right derivate of f at x_0 is*

$$D^+f(x_0) = \limsup_{x \to x_0+} \frac{f(x) - f(x_0)}{x - x_0};$$

(3.1.5) *the lower right derivate of f at x_0 is*

$$D_+f(x_0) = \liminf_{x \to x_0+} \frac{f(x) - f(x_0)}{x - x_0}.$$

If $a < x_0$,

(3.1.6) *the upper left derivate of f at x_0 is*

$$D^-f(x_0) = \limsup_{x \to x_0-} \frac{f(x) - f(x_0)}{x - x_0};$$

(3.1.7) *the lower left derivate of f at x_0 is*

$$D_-f(x_0) = \liminf_{x \to x_0-} \frac{f(x) - f(x_0)}{x - x_0}.$$

The four one-sided limits [(3.1.4)–(3.1.7)] are called the *Dini derivatives* or *Dini derivates* of f at the point $x_0 \in [a, b]$. Note that we allow the possibility of infinite values for any of the derivates. Although we shall be concerned primarily with functions defined on intervals, the above definitions can be applied to functions defined on an arbitrary set E, where only difference quotients using points in E are considered. In Chapter 1 (Definition 1.2.1), we defined derived numbers; the reader will note that the Dini derivatives are derived numbers. In fact, some authors use the term derived numbers for the quantities we have defined as the Dini derivatives.

It is easy to verify the following properties of Dini derivatives.

Theorem 3.1.8 [M2, pp. 189–191]. *Let f be defined and finite on the interval $[a, b]$. Then*

(3.1.9)

$$\overline{D}f(a) = D^+f(a),$$
$$\underline{D}f(a) = D_+f(a),$$
$$\overline{D}f(b) = D^-f(b),$$
$$\underline{D}f(b) = D_-f(b).$$

(3.1.10) If $x_0 \in (a, b)$,

$$\overline{D}f(x_0) = \sup\{D^+f(x_0), D^-f(x_0)\},$$
$$\underline{D}f(x_0) = \inf\{D_+f(x_0), D_-f(x_0)\}.$$
$$D^+(-f(x)) = -D_+f(x),$$
$$D_+(-f(x)) = -D^+f(x),$$

$$D^+f(-x) = -D_-f(x),$$
$$D_+f(-x) = -D^-f(x),$$
$$D^+(-f(-x)) = D^-f(x),$$

(3.1.11) $\qquad\qquad D_+(-f(-x)) = D_-f(x).$

In (3.1.11), we may interchange the superscripts $+$ and $-$ on the letter D, as well as the subscripts.

The following simple examples illustrate the calculation of Dini derivatives.

EXAMPLE 3.1.12. Let $f(x) = |x|$. Then $D^+f(0) = D_+f(0) = 1$ and $D^-f(0) = D_-f(0) = -1$.

EXAMPLE 3.1.13. Let

$$f(x) = \begin{cases} 0, & x = 0 \\ x \sin\dfrac{1}{x}, & x \neq 0. \end{cases}$$

Then $D_+f(0) = D_-f(0) = -1$ and $D^+f(0) = D^-f(0) = 1$.

EXAMPLE 3.1.14. Let $a \leq b, c \leq d$, and

$$f(x) = \begin{cases} ax\left(\sin\dfrac{1}{x}\right)^2 + bx\left(\cos\dfrac{1}{x}\right)^2, & x > 0 \\ 0, & x = 0 \\ cx\left(\sin\dfrac{1}{x}\right)^2 + dx\left(\cos\dfrac{1}{x}\right)^2, & x < 0. \end{cases}$$

Then $D_+f(0) = a \leq b = D^+f(0)$ and $D_-f(0) = c \leq d = D^-f(0)$. **Note:** For $h > 0$,

$$\frac{f(0 + h) - f(0)}{h} = a + (b - a)\left(\cos\frac{1}{h^2}\right).$$

Thus,

$$a \leq \frac{f(0 + h) - f(0)}{h} \leq b.$$

For $h = 1/\pi, 1/2\pi, \ldots$,

$$\frac{f(0 + h) - f(0)}{h} = b.$$

For $h = 2/\pi, 2/3\pi, \dots$,

$$\frac{f(0 + h) - f(0)}{h} = a.$$

Definition 3.1.15. Let f be defined and finite on the interval $[a, b]$ and let $x_0 \in [a, b]$. Then

(a) if $x_0 < b$ and $D^+ f(x_0) = D_+ f(x_0)$, their common value is called the *right derivative* of f at x_0 and is denoted $f'_+(x_0)$;
(b) if $a < x_0$ and $D^- f(x_0) = D_- f(x_0)$, their common value is called the *left derivative* of f at x_0 and is denoted $f'_-(x_0)$;
(c) if $\overline{D}f(x_0) = \underline{D}f(x_0)$, their common value is the derivative of f at x_0 and is denoted $f'(x_0)$.

§3.2. Simple Properties of Derivatives

Theorem 3.2.1 [M2, p. 193]. *Let f be defined and finite on $[a, b]$ and let $x_0 \in [a, b]$. If $\overline{D}f(x_0)$ and $\underline{D}f(x_0)$ are both finite, then f is continuous at x_0.*

Proof. Let M be the greater of $|\overline{D}f(x_0)|$ and $|\underline{D}f(x_0)|$. Then there exists $\delta > 0$ such that

$$-M' \le \frac{f(x) - f(x_0)}{x - x_0} \le M'$$

(M' depends only on M), if $x \ne x_0$ is in $[a, b] \cap N_\delta(x_0)$. Then

$$|f(x) - f(x_0)| \le M'|x - x_0|.$$

For $\varepsilon > 0$ and $\gamma = \min(\delta, \varepsilon/M')$,

$$|f(x) - f(x_0)| < \varepsilon$$

if $x \in [a, b]$ and $|x - x_0| < \gamma$.

Theorem 3.2.2 [H2, p. 365]. *Let f be finite and continuous on the closed interval $[a, b]$. Suppose one of the four Dini derivatives is zero at every point of $[a, b]$, with the exception of a denumerable set; then f is constant on the interval $[a, b]$.*

Proof. We will prove the theorem for the case $D^+ f(x) = 0$ for all $x \in [a, b] - E$, E a denumerable set. Suppose that f is not constant on $[a, b]$; then there exists a point $x_1 \in (a, b]$ such that $f(x_1) - f(a) = p \ne 0$. Suppose $p > 0$. Define the function

$$\phi_k(x) = f(x) - f(a) - k(x - a).$$

Then $\phi_k(a) = 0$ and $\phi_k(x_1) = p - k(x_1 - a)$. Fix q such that $0 < q < p$. If $k < (p - q)/(x_1 - a) = K$, then $\phi_k(x_1) > q$. Now ϕ_k is continuous in the interval $[a, b]$. Let

$$A = \{x \in (0, x_1): \phi_k(x) \le q\}.$$

Let $\xi = \sup A$. Then $\xi < x_1$ and $\phi_k(\xi) = q$. If $0 < h \le x_1 - \xi$, then $\phi_k(\xi + h) > q$. Thus,

$$\frac{\phi_k(\xi + h) - \phi_k(\xi)}{h}$$

is positive; hence, $D^+ \phi_k(\xi) \ge 0$. If $\xi \notin E$, then

$$D^+ \phi_k(\xi) = D^+ f(x) - k = -k.$$

Therefore, ξ must be in E. □

Since the number q is fixed, ξ depends only on k. Corresponding to a given value of ξ, there is only one value of k, for

$$\phi_k(\xi) - \phi_{k'}(\xi) = (k - k')(\xi - a)$$

which is not zero unless $k = k'$. Thus, for a given k, the corresponding values of ξ (all in E) must be denumerable. Therefore, to each value of k, in the interval $(a, K - b)$ there is a denumerable set of values of ξ. This would imply that the interval $(a, K - b)$ is itself denumerable, which is not true. Thus, at no point $x \in (a, b]$ can $f(x) - f(a)$ be positive. By considering the function

$$f(x) - f(a) + k(x - a)$$

we can show that $f(x) - f(a)$ cannot be negative for any $x \in (a, b]$. Hence, $f(x) = f(a)$ for every $x \in (a, b]$. The case in which one of the other three Dini derivatives is zero, except on a denumerable set, can be proved in a similar manner.

Theorem 3.2.3 [H2, p. 366]. *If two continuous finite functions are defined on the closed interval $[a, b]$, and if, for one of the four Dini derivatives, this derivative for the two functions has finite equal values at every point of $[a, b]$, except for a denumerable set, then the two functions differ from each other only by a constant on the entire interval.*

Proof. Suppose f_1 and f_2 are two continuous functions such that

$$D^+ f_1(x) = D^+ f_2(x)$$

at every point $x \in [a, b] - E$, E a denumerable set. Let

$$f(x) = f_1(x) - f_2(x).$$

Suppose $\varepsilon > 0$. Then for any $x \notin E$,

$$\frac{f_1(x + h) - f_1(x)}{h} > D^+ f_1(x) - \varepsilon$$

for sufficiently small values of h. Similarly, for sufficiently small h,

$$\frac{f_2(x + h) - f_2(x)}{h} < D^+ f_2(x) + \varepsilon.$$

Thus [since $D^+ f_1(x) = D^+ f_2(x)$ for any $x \notin E$],

$$\frac{f(x + h) - f(x)}{h} > -2\varepsilon$$

for such h. Hence, $D^+ f(x) > -2\varepsilon$. Because ε is arbitrary, we conclude that $D^+ f(x) \geq 0$. By interchanging $f_1(x)$ and $f_2(x)$, we get $D^+(-f(x)) \geq 0$. Then, by Exercise 1, we have $f(x) - f(a) \geq 0$ and $f(a) - f(x) \geq 0$ for all $x \in (a, b]$. Therefore, $f(x) = f(a)$ everywhere and the theorem is proved.   □

§3.3. Ruziewicz's Example

We recall from elementary calculus the following result:

If f and g are two continuous functions defined on the interval $[a, b]$ and have equal finite derivatives in (a, b), then $f - g$ is constant on $[a, b]$.

Theorem 3.2.3 is a natural extension of this result. But if we remove the requirement that the derivatives be finite, then the result no longer holds, as seen in the following example from Ruziewicz [R6].

We will construct an infinite number of functions which have the same derivative (not finite everywhere) in an interval, but the difference of any two is not a constant. Let s be a real number greater than or equal to 1. Divide the segment $[0, s]$ into three parts so that the center of the middle segment is the center of the segment $[0, s]$ and the middle segment has length $\frac{1}{3}$. The three segments obtained are

$$\left[0, \frac{s}{2} - \frac{1}{6} \right], \quad \left[\frac{s}{2} - \frac{1}{6}, \frac{s}{2} + \frac{1}{6} \right], \quad \left[\frac{s}{2} + \frac{1}{6}, s \right].$$

Associated with the middle segment $[s/2 - 1/6, s/2 + 1/6]$, there is a half-circle in the plane (lying above the x axis) whose diameter is $\frac{1}{3}$, the length of the segment.

At the second step of the process, we consider the first and third intervals from the division above; i.e.,

$$\left[0, \frac{s}{2} - \frac{1}{6} \right], \quad \left[\frac{s}{2} + \frac{1}{6}, s \right]$$

each of which has length $\geq \frac{1}{3}$. We divide each of these segments into three parts, the middle part of each having length $1/3^2$ and the center of the middle segment again being the center of the segment which is being divided. The two middle segments obtained at this step are

$$\left[\frac{s}{4} - \frac{5}{36}, \frac{s}{4} - \frac{1}{36}\right], \quad \left[\frac{3s}{4} + \frac{1}{36}, \frac{3s}{4} + \frac{5}{36}\right].$$

We associate with each of the above segments the half-circle in the plane with the segment as its diameter; there are two such half-circles. The four segments which are not middle segments will each have length $\geq 1/3^2$.

The next step is to divide each of the four segments, which are not middle segments of the previous division, into three parts, the center of the middle parts being the center of the segment which is being divided and the middle parts each having length $1/3^3$. In each of these four middle segments we construct a half-circle having the middle segment as its diameter; there are four such half-circles obtained at this step.

Repeating this procedure indefinitely, at the nth step, we have 2^{n-1} half-circles of diameter $1/3^n$. We thus create an infinite denumerable sequence of half-circles. The accumulation points of the set of points lying on the half-circles form a perfect nondense set P_s in the interval $[0, s]$. Let Q_s be the set of all those points (in the plane) lying on the arcs of the half-circles determined by the process described above.

The set Q_s can be thought of as the graph of a function

$$y = f_s(x)$$

determined by the number s and continuous for $0 \leq x \leq 1$. Let

$$F_s(t) = \int_0^t f_s(x)\,dx \quad \text{for } 0 \leq t \leq s.$$

Then $F_s(t)$ is a continuous function which is increasing on the interval $[0, s]$. Using the geometric significance of the integral, the number $F_s(s)$ is the sum of the areas of all the half-circles we have constructed:

$$F_s(s) = \frac{\pi}{8}\left(\frac{1}{3^2} + \frac{2}{3^4} + \frac{2^2}{3^6} + \cdots\right) = \frac{\pi}{56}.$$

Let $\phi_s(u)$ be the inverse function of $F_s(t)$. Since $F_s(t)$ is continuous and increasing, then $\phi_s(u)$ is continuous and increasing for $0 \leq u \leq F_s(s)$, i.e., on the interval $[0, \pi/56]$. In addition, $0 \leq \phi_s(u) \leq s$ for $0 \leq u \leq \pi/56$.

Now we evaluate the derivative $\phi_s'(u)$ for $u \in [0, \pi/56]$. There are two cases: First, suppose $\phi_s(u) = t_s \notin P_s$. Then $f_s(t_s) \neq 0$. By continuity of $f_s(t)$, $F_s'(t_s) = f_s(t_s)$. Thus,

(3.3.1) $$\phi_s'(u) = \frac{1}{F_s'(t_s)} = \frac{1}{f_s(t_s)}.$$

Since $t_s = \phi_s(u)$, we have $F_s(t_s) = u$ for all $s \geq 1$, which gives

(3.3.2) $$F_s(t_s) = F_1(t_1).$$

Because of the construction of the functions f_s and the geometric significance of the integral, we conclude

(3.3.3) $$f_s(t_s) = f_1(t_1).$$

From (3.3.1) and (3.3.3), we have

(3.3.4) $$\phi'_s(u) = \phi'_1(u).$$

Now we consider the second case. Suppose $\phi_s(u) = t_s \in P_s$. Then we have $f_s(t_s) = 0$; hence, $F'_s(t_s) = 0$. Since ϕ_s is the inverse of F_s and ϕ_s is increasing in $[0, \pi/56]$, we conclude that

$$\phi'_s(u) = +\infty,$$

so that (3.3.4) is also true in this case. Thus, (3.3.4) holds for all $u \in [0, \pi/56]$. Therefore, all the functions ϕ_s, $s \geq 1$, have the same derivative in the interval $[0, \pi/56]$. We will now show that the difference of any two of these functions is not a constant.

Since $F_s(s) = \pi/56$ and $F_s(0) = 0$ for every s, then $\phi_s(\pi/56) = s$ and $\phi_s(0) = 0$ for every s. Thus, if $s \neq s'$,

$$\phi_s(0) - \phi_{s'}(0) = 0,$$

but

$$\phi_s\left(\frac{\pi}{56}\right) - \phi_{s'}\left(\frac{\pi}{56}\right) = s - s' \neq 0.$$

Therefore, the difference of any two of the functions ϕ_s and $\phi_{s'}$ is not constant in the interval $[0, \pi/56]$.

§3.4. Further Properties of Derivatives

Theorem 3.4.1. *Let f be continuous in $[a, b]$ and differentiable at $x = a$ and $x = b$. Further, suppose the four derivatives D^+f, D_+f, D^-f, D_-f are finite in (a, b). Then if $f'(a) = f'(b)$, there exists $\xi_1 \in (a, b)$ such that*

(3.4.2) $$D^+f(\xi_1) \leq \frac{f(\xi_1) - f(a)}{\xi_1 - a} \leq D_-f(\xi_1);$$

or there exists $\xi_2 \in (a, b)$ such that

(3.4.3) $$D^-f(\xi_2) \leq \frac{f(\xi_2) - f(a)}{\xi_2 - a} \leq D_+f(\xi_2).$$

Proof. Suppose $f'(a) = f'(b) = 0$ [otherwise consider the function $f(x) - xf'(a)$]. Define

$$g(x) = \begin{cases} f'(a) = 0, & x = a \\ \dfrac{f(x) - f(a)}{x - a}, & x \in (a, b]. \end{cases}$$

Now g is continuous on $[a, b]$ and

$$g'(b) = \frac{-g(b)}{b - a}.$$

If $g(b) > 0$, then $g'(b) < 0$. Hence, g is decreasing at b, whereas $g(a) = 0$. Since g is continuous on $[a, b]$, it attains its maximum at some $\xi_1 \in (a, b)$. Hence,

$$D^+ g(\xi_1) \le 0, \qquad D_- g(\xi_1) \ge 0.$$

But

$$D^+ g(\xi_1) = \frac{D^+ f(\xi_1)}{\xi_1 - a} - \frac{f(\xi_1) - f(a)}{(\xi_1 - a)^2}$$

and

$$D_- g(\xi_1) = \frac{D_- f(\xi_1)}{\xi_1 - a} - \frac{f(\xi_1) - f(a)}{(\xi_1 - a)^2}.$$

Inequality (3.4.2) then follows. If, however, $g(b) < 0$, then $g'(b) > 0$. Thus, g is increasing at b, whereas $g(a) = 0$. Hence, g attains a minimum value at some $\xi_2 \in (a, b)$. Therefore,

$$D_+ g(\xi_2) \ge 0, \qquad D^- g(\xi_2) \le 0,$$

and we get (3.4.3). If $g(b) = 0$, since g is continuous and $g(a) = 0$, g attains a maximum at some ξ_1 or a minimum at some ξ_2 and again we get the desired result.   □

A characterization of the monotonicity of a function by using Dini derivatives is given by the following theorem.

Theorem 3.4.4 [M2, p. 200]. *If f is defined and finite on $[a, b]$ and*

(a) $\limsup_{x \to x_0-} f(x) \le f(x_0) \, (a < x_0 \le b)$
(b) $\liminf_{x \to x_0+} f(x) \ge f(x_0) \, (a \le x_0 < b)$

and one of the following two conditions holds:

(c) $D^+ f(x) \ge 0$ *for all* $x \in [a, b]$ *except at most a denumerable subset* E_0
(d) $D^- f(x) \ge 0$ *for all* $x \in [a, b]$ *except at most a denumerable subset* E_0

then f is increasing on $[a, b]$.

Proof. Let us assume that $D^+ f(x) > 0$ except for $x \in E_0$. If the result is not true, then there exists α and β such that $a \leq \alpha < \beta \leq b$ and $f(\alpha) > f(\beta)$. Since E_0 is at most denumerable, the values of $f(x)$ for $x \in E_0$ cannot fill up the interval $(f(\beta), f(\alpha))$. Thus, there exists μ such that

$$f(\beta) < \mu < f(\alpha),$$

and for all $x \in E_0$, $f(x) \neq \mu$. Since $f(\alpha) > \mu$, the set of points in $[\alpha, \beta]$ at which $f(x) \geq \mu$ is nonempty. Let ξ be the sup of such x. We claim that $f(\xi) = \mu$.

If not, suppose $f(\xi) > \mu$. Since $f(\beta) < \mu$, $\xi \neq \beta$. By definition of ξ, we have $f(x) < \mu$ for $x \in (\xi, \beta]$. Thus,

$$\liminf_{x \to \xi+} f(x) \leq \mu < f(\xi),$$

contradicting hypothesis (b).

Now let $f(\xi) < \mu$. Since $f(\alpha) > \mu$, $\xi \neq \alpha$. By definition of ξ, for every $\delta > 0$, there exists $x \in [\alpha, \beta] \cap (\xi - \delta, \xi]$ such that $f(x) \geq \mu$. But $f(\xi) < \mu$ and, thus, the required x is not ξ. Hence, $x \in [\alpha, \xi) \cap N_\delta(\xi)$. Thus,

$$\sup_{x \in N_\delta(\xi) \cap [\alpha, \xi)} f(x) \geq \mu$$

and

$$\limsup_{x \to \xi-} f(x) \geq \mu > f(\xi),$$

contradicting (a). Hence, $f(\xi) = \mu$.

Now

$$\frac{f(x) - f(\xi)}{x - \xi} < 0, \quad x \in (\xi, \beta].$$

Thus, $D^+ f(\xi) \leq 0$. Since $f(\xi) = \mu$ and $f(x) \neq \mu$, if $x \in E_0$, then $\xi \notin E_0$. This gives a contradiction. This establishes the result for $D^+ f(x)$ positive except on E_0.

We now suppose that (a), (b), and (c) hold and we apply the above results to the function $f(x) + \varepsilon x$ ($\varepsilon > 0$) which has an upper derivate

$$D^+(f(x) + \varepsilon x) = D^+ f(x) + \varepsilon \geq \varepsilon > 0$$

except perhaps on E_0. This would lead to the conclusion that $f(x) + \varepsilon x$ is increasing, i.e., $a < \alpha < \beta \leq b$ implies

$$f(\beta) - f(\alpha) \geq -\varepsilon(\beta - \alpha).$$

Since ε is arbitrary, f is increasing.

If (a), (b), and (d) hold, we consider the function

$$g(x) = -f(-x), \quad -b \leq x \leq -a,$$

and let E_1 be the denumerable set of points x such that $-x \in E_0$. Then g satisfies (a), (b), and (c). Thus, g is increasing; therefore, $f(x) = -g(-x)$ is increasing. $\qquad \square$

The following theorem illustrates the connection between the maximum slope of all the chords connecting points on the graph of a function and its four Dini derivatives.

Theorem 3.4.5 [M2, p. 206]. *Let f be defined and continuous on $[a, b]$. Then*

$$\sup_{x_1, x_2 \in [a,b]} \frac{f(x_1) - f(x_2)}{x_1 - x_2} = \sup_{x \in [a,b]} D^+ f(x)$$

$$= \sup_{x \in [a,b]} D_+ f(x)$$

$$= \sup_{x \in [a,b]} D^- f(x)$$

$$= \sup_{x \in [a,b]} D_- f(x).$$

A similar equality holds for the inf.

Proof. Let

$$M = \sup_{x_1, x_2 \in [a,b]} \frac{f(x_1) - f(x_2)}{x_1 - x_2}.$$

Clearly, M is an upper bound on $[a, b]$ for all four functions $D^+ f$, $D_+ f$, $D^- f$, $D_- f$. Let

$$m = \sup_{x \in [a,b]} D_+ f(x).$$

Then $m \leq M$. Suppose $m < M$. Then there exists a k such that $m < k < M$ and

$$g(x) = kx - f(x)$$

is continuous on $[a, b]$. Since

$$D^+ g(x) = k - D_+ f(x) > 0,$$

g is increasing on $[a, b]$. For all $x_1, x_2 \in [a, b]$, $x_2 > x_1$, we have

$$0 \leq g(x_2) - g(x_1) = k(x_2 - x_1) - [f(x_2) - f(x_1)]$$

or

$$\frac{f(x_2) - f(x_1)}{x_2 - x_1} \leq k.$$

This is also true if $x_2 < x_1$. Thus, $k < M$ is an upper bound and we get a contradiction. Similarly,

$$\sup_{x \in [a,b]} D_- f(x) = M.$$

Since $D^+ f(x) \geq D_+ f(x)$, we have

$$M \geq \sup D^+ f(x) \geq \sup D_+ f(x) = M.$$

Thus,

$$\sup D^+ f(x) = M$$

and, similarly,

$$D^- f(x) = M.$$

Replacing f by $-f$ and applying (3.1.11) yields the desired result for the inf. $\square$

As a consequence of Theorem 3.4.5, we have the following result.

Theorem 3.4.6. *If any one of the four Dini derivatives of a continuous function f is continuous at the point x_0, so are the other three. In this case, all four Dini derivatives are equal and f is differentiable at x_0.*

The next theorem gives a sufficient condition in terms of the Dini derivatives for a function to be Lipschitz.

Theorem 3.4.7 [M2, p. 207]. *If f is defined and finite on $[a,b]$ and its Dini derivatives are bounded, then f satisfies a Lispschitz condition on $[a,b]$.*

Proof. Let M be an upper bound for all the Dini derivatives for all $x \in [a,b]$. Then by Theorem 3.4.5, for any $x_1, x_2 \in [a,b]$,

$$\left| \frac{f(x_1) - f(x_2)}{x_1 - x_2} \right| \leq M$$

and the result follows. $\square$

Theorem 3.4.8 [S1, p. 261]. *Let f be a finite function. Then each of the following sets is at most denumerable:*

(a) *the set of points at which f attains a strict maximum or minimum;*
(b) *the set of points x at which*

$$\limsup_{t \to x} f(t) > \limsup_{t \to x+} f(t)$$

   or

$$\liminf_{t \to x} f(t) < \liminf_{t \to x+} f(t)$$

Proof. (a) Let A be the set of points where f attains a strict maximum. For each positive integer n, let

$$A_n = \left\{ x : f(t) < f(x) \text{ holds } \forall t (\neq x) \in \left(x - \frac{1}{n}, x + \frac{1}{n} \right) \right\}.$$

Clearly, A_n does not contain any of its accumulation points. Hence, each A_n is at most denumerable and

$$A = \bigcup_n A_n$$

is at most denumerable.

(b) Let

$$B = \left\{ x: \limsup_{t \to x} f(t) > \limsup_{t \to x+} f(t) \right\}.$$

For each pair of integers p and q, let

$$B_{pq} = \left\{ x: \limsup_{t \to x} f(t) > \frac{p}{q} > \limsup_{t \to x+} f(t) \right\}.$$

Each point of B_{pq} is an isolated point of B_{pq} from the right. Each B_{pq} is then at most denumerable and, hence, so is

$$B = \bigcup_{p,q \in N} B_{pq} \qquad\qquad \square$$

§3.5. The Denjoy–Saks–Young Theorem

In this section we characterize the Dini derivatives of an arbitrary function, culminating in the Denjoy–Saks–Young Theorem (3.5.5).

Lemma 3.5.1 [H2, p. 391]. *For every real-valued finite function f and any fixed real number k, the set of points*

$$\{x: D_+ f(x) \geq k \text{ and } D^- f(x) < k\}$$

is denumerable. The same is true for the sets

$$\{x: D_- f(x) \geq k \text{ and } D^+ f(x) < k\},$$

$$\{x: D^+ f(x) \leq k \text{ and } D_- f(x) > k\},$$

$$\{x: D^- f(x) \leq k \text{ and } D_+ f(x) > k\}.$$

Proof. Let E be the set of points at which $D_+ f(x) \geq k$. Let $\{\varepsilon_n\}$ and $\{\eta_n\}$ be two monotone sequences of positive numbers which converge to zero. If $\xi \in E$ and h is a sufficiently small positive number

$$(3.5.2) \qquad \frac{f(\xi + h) - f(\xi)}{h} \geq k - \eta_n.$$

Let $\bar{h}$ be the upper bound of those h's which satisfy (3.5.2). Let $\bar{\xi} = \xi + \frac{1}{2}\bar{h}$.

Then for every $x \in (\xi, \bar{\xi})$,

$$\frac{f(x) - f(\xi)}{x - \xi} \geq k - \eta_n.$$

Let $\delta_\xi = \min\{\varepsilon_n, \bar{\xi} - \xi\}$. Then for $x \in (\xi, \xi + \delta_\xi)$,

$$\frac{f(x) - f(\xi)}{x - \xi} \geq k - \eta_n$$

and

$$|\xi - x| \leq \varepsilon_n$$

are true. Let Δ_n be the set of all such intervals $(\xi, \xi + \delta_\xi)$ for all $\xi \in E$, for a given value of n. Let $\Delta = \{\Delta_n : n \in N\}$. Let H be the set of points in E that are not interior points of any interval of Δ. Then H is at most denumerable. If $\xi' \in E - H$ and n is any positive integer, then ξ' is in the interior of some interval (ξ, η) in Δ_n. Thus,

$$\frac{f(\xi') - f(\xi)}{\xi' - \xi} \geq k - \eta_n.$$

As $n \to \infty$, $\xi \to \xi'$ and $\eta_n \to 0$. Hence, $D^- f(\xi') \geq k$. This result holds for every $\xi' \in E - H$, and since H is denumerable, we have proved the result for the first set. The proof for the second set is similar. Substituting $-f$ for f and $-k$ for k, we can prove the results for the last two sets. $\qquad\square$

Corollary 3.5.3 [H2, p. 392]. *The following sets of points are denumerable:*

(a) $\{x : D_+ f(x) = +\infty \text{ and } D^- f(x) = -\infty \text{ or is finite}\}$;
(b) $\{x : D_- f(x) = +\infty \text{ and } D^+ f(x) = -\infty \text{ or is finite}\}$;
(c) $\{x : D^+ f(x) = -\infty \text{ and } D_- f(x) = +\infty \text{ or is finite}\}$;
(d) $\{x : D^- f(x) = -\infty \text{ and } D_+ f(x) = +\infty \text{ or is finite}\}$.

Proof. The set of points at which $D_+ f(x) = +\infty$ and $D^- f(x) < k$ is a subset of the denumerable set at which $D_+ f(x) \geq k$ and $D^- f(x) < k$. Let $\{k_n\}$ be a sequence of positive numbers such that $k_n \to \infty$. Then the set for which $D_+ f(x) = +\infty$ and $D^- f(x) < +\infty$ belongs to one of the sets for which $D^+ f(x) = +\infty$ and $D^- f(x) < k_n$. Thus, the set described in (a) is denumerable. The proofs for the other three sets are similar. $\qquad\square$

Theorem 3.5.4 [H2, p. 392]. *For every real-valued finite function f,*

$$D^+ f(x) \geq D_- f(x) \qquad and \qquad D^- f(x) \geq D_+ f(x)$$

except at points of a denumerable set.

Proof. Suppose x_0 is a point at which $D^+ f(x_0) = \alpha$ and $D_- f(x_0) = \beta$, where $\alpha < \beta$; β may be $+\infty$. Let k be a rational number such that $\alpha < k < \beta$. Then

x_0 belongs to the set of points

$$E_k = \{x: D^+f(x) < k \text{ and } D_-f(x) > k\}.$$

By Theorem 3.5.1, E_k is denumerable. Any point x at which $D^+f(x) < D_-f(x)$ belongs to some E_k. If $D^+f(x)$ is finite or $-\infty$ and $D_-f(x) = +\infty$, then x belongs to the denumerable set E_∞. Since the set

$$\left[\bigcup_{k \in Q} E_k\right] \cup E_\infty$$

is denumerable, the set of points at which $D^+f(x) < D_-f(x)$ is denumerable. The proof that the set

$$\{x: D^-f(x) < D_+f(x)\}$$

is denumerable is similar.   □

Theorem 3.5.5 (Denjoy–Saks–Young Theorem) [R3, p. 18]. *Let f be a finite real-valued function defined on an interval $[a, b]$. Then at every point of $[a, b]$, except for a set of measure zero, either*

(a) *there is a finite derivative; or*
(b) *D^+f and D_-f are finite and equal, $D^-f = +\infty$, and $D_+f = -\infty$; or*
(c) *D^-f and D_+f are finite and equal, $D^+f = +\infty$, and $D_-f = -\infty$; or*
(d) *$D^+f = D^-f = +\infty$ and $D_+f = D_-f = -\infty$.*

Proof. We will show that at almost all points where $D_-f(x) \neq -\infty$, $D_-f(x)$ is finite, and $D_-f(x) = D^+f(x)$. The theorem then follows by applying this result and (3.1.11) to the functions $-f$, $f(-x)$, and $-f(-x)$. Suppose that f is defined on the interval $[a, b]$. Let E be the set of points $x \in [a, b]$ where $D_-f(x) \neq -\infty$. For $n = 0, 1, 2, \ldots$ and r a rational number in $[a, b]$, let

$$E_{n,r} = \left\{x > r: \frac{f(x) - f(\xi)}{x - \xi} > -n \text{ for all } \xi \in (r, x)\right\}.$$

Then

$$E = \bigcup_{n=0}^{\infty} \bigcup_{r \in [a,b] \cap Q} E_{n,r}.$$

Since the union of a denumerable collection of sets of measure zero is itself of measure zero, it will suffice to show that each $E_{n,r}$ is of measure zero. In addition, by replacing $f(x)$ by $f(x - r) + nx$, we consider only the set $E_{0,0} = E_0$. Exclude those points of E_0 which are not points of density of E_0 (a set of measure zero). We also exclude those points of E_0 where f does not possess a finite derivative with respect to the set E_0; again we exclude only a set of measure zero because $f|_{E_0}$ is monotonic and thus has a finite derivative a.e. on E_0. On the points of E_0 which remain, consider the ratio

$$\frac{f(x') - f(x)}{x' - x}.$$

Let $x' \to x$ without leaving the set E_0; call this limit $f'_{E_0}(x)$ (such a limit exists because we have excluded from consideration all points where a finite derivative does not exist). If $x' \notin E_0$ but is sufficiently close to x, because of the density hypothesis we can replace x' by a $\xi > x'$ such that $\xi \in E_0$ and the difference $\xi - x'$ is as close to $x' - x$ as we desire. By definition of the set E_0, $f(\xi) \geq f(x')$. Therefore, the numerator of the above ratio does not decrease when we replace x' by ξ and the new denominator is close to the previous one. Thus, we have

$$D_- f(x) \geq f'_{E_0}(x) \geq D^+ f(x).$$

But by definition of $f'_{E_0}(x)$, we have

$$D_- f(x) \leq f'_{E_0}(x) \qquad \text{and} \qquad D^+ f(x) \geq f'_{E_0}(x).$$

Thus, equality must hold and we have proved the desired result. $\square$

The following corollaries are immediate consequences of Theorem 3.5.5.

Corollary 3.5.6 [H2, p. 400]. *If a function has finite Dini derivatives except at points of a set of measure zero, it has a finite derivative almost everywhere.*

Corollary 3.5.7 [H2, p. 400]. *A monotone function has a finite derivative except on a set of measure zero (cf. Theorem 1.2.8).*

Corollary 3.5.8 [H2, p. 401]. *The points at which a continuous function has an infinite derivative (of fixed sign) form a set of measure zero.*

Remark. We began the chapter with the assumption that f is a finite function. If we modify that assumption to allow f to be infinite on a set of measure zero, Theorem 3.5.5 will still hold. In determining the value of the Dini derivatives, we will ignore pairs of points where the function has the same infinite value (to avoid expressions of the form $\infty - \infty$). At the points where $f(x) = +\infty$, we have

$$D^+ f(x) = D_+ f(x) = -\infty$$

and

$$D^- f(x) = D_- f(x) = +\infty;$$

whereas at points where $f(x) = -\infty$, we have

$$D^+ f(x) = D_+ f(x) = +\infty$$

and

$$D^- f(x) = D_- f(x) = -\infty.$$

Thus, it appears that the set of points where the upper derivative on one side is less than the lower derivative on the other side contains the set of points where the function is infinite.

§3.6. Measurability of Dini Derivatives

The next important result we shall prove is that the Dini derivatives of a measurable function are measurable (Theorem 3.6.5). This theorem and the lemmas preceding it are due to Banach [B1].

Lemma 3.6.1. *Let f be a function of a real variable (measurable or non-measurable) and δ a positive real number. Define*

$$F(x) = \sup_{0 \le h \le \delta} f(x + h).$$

Then the discontinuities (if any) of F are ordinary discontinuities; hence, F is of Baire class ≤1.

Proof. Let

$$B(\alpha, \beta) = \sup_{\alpha \le x \le \beta} f(x).$$

Let x_0 be a fixed real number and x any number such that $0 < x - x_0 < \delta$. The interval $(x, x + \delta)$ is the union of the two intervals $(x, x_0 + \delta)$ and $(x_0 + \delta, x + \delta)$. Thus,

$$F(x) = B(x, x + \delta) = \max\{B(x, x_0 + \delta), B(x_0 + \delta, x + \delta)\}.$$

When x is decreased, $B(x, x + \delta)$ will not decrease and $B(x_0 + \delta, x + \delta)$ will not increase. Therefore, $B(x, x_0 + \delta)$ and $B(x_0 + \delta, x + \delta)$ will have a limit as x decreases to x_0. Thus,

$$\lim_{x \to x_0+} F(x)$$

exists. In a similar manner, it can be shown that the limit from the left also exists. Thus, the lemma is proved. ☐

Corollary 3.6.2. *For $0 < a < b$, let*

$$F(x, a, b) = \sup_{a \le h \le b} f(x + h).$$

Then F is a function (in x) of Baire class ≤1.

Proof. We observe that $F(x, a, b) = B(x + a, x + b)$. By Lemma 3.6.1,

$$\phi(x) = B(x, x + h - a)$$

is of class ≤1. Therefore, $F(x, a, b) = \phi(x + a)$ is of class ≤1. ☐

Corollary 3.6.3. *Let f be a function of Baire class $\alpha > 0$. For $0 < a < b$, let*

$$\phi(x, a, b) = \sup_{a \le h \le b} \{f(x + h) - f(x)\}.$$

Then ϕ is function of Baire class ≤α.

Proof. Since $\phi(x, a, b) = F(x, a, b) - f(x)$, we can use Corollary 3.6.2 and the fact that the difference of two functions of class $\leq \alpha$ is a function of class $\leq \alpha$ to obtain the result. $\qquad\qquad\square$

Corollary 3.6.4. *Let f be a measurable function. Then ϕ of the previous corollary is a measurable function.*

Theorem 3.6.5. *The Dini derivatives of a measurable function are measurable.*

Proof. We shall prove the result for D^+f, the other cases being similar. First, suppose f is bounded. Let $a < b$ be two positive numbers. Let

$$\Phi(x, a, b) = \sup_{a \leq h \leq b} \frac{f(x + h) - f(x)}{h}$$

and

$$\phi(x, a, b) = \sup_{a \leq h \leq b} \{f(x + h) - f(x)\}.$$

Let n be any positive integer and

(3.6.6) $$a_k = a + \frac{k}{n}(b - a), \quad k = 0, 1, \ldots, n.$$

Then

(3.6.7) $$\Phi(x, a, b) = \max_{k \leq n} \phi(x, a_{k-1}, a_k).$$

We have, for $0 < \alpha < \beta$ and $\alpha \leq h \leq \beta$,

$$f(x + h) - f(x) \leq \phi(x, \alpha, \beta)$$

and

$$\frac{1}{\beta} \leq \frac{1}{h} \leq \frac{1}{\alpha}.$$

Hence, if $\alpha \leq h \leq \beta$, for $\phi(x, \alpha, \beta) \geq 0$ we have

$$\frac{f(x + h) - f(x)}{h} \leq \frac{\phi(x, \alpha, \beta)}{\alpha}$$

and for $\phi(x, \alpha, \beta) \leq 0$,

$$\frac{f(x + h) - f(x)}{h} \leq \frac{\phi(x, \alpha, \beta)}{\beta}.$$

Consequently,

(3.6.8)
$$\Phi(x, \alpha, \beta) \leq \frac{\phi(x, \alpha, \beta)}{\alpha} \quad \text{if } \phi(x, \alpha, \beta) \geq 0,$$

$$\Phi(x, \alpha, \beta) \leq \frac{\phi(x, \alpha, \beta)}{\beta} \quad \text{if } \phi(x, \alpha, \beta) \leq 0.$$

Thus,

$$f(x + h) - f(x) \leq h\Phi(x, \alpha, \beta)$$

if $\alpha \leq h \leq \beta$. It follows that

(3.6.9)
$$\phi(x, \alpha, \beta) \leq \beta\Phi(x, \alpha, \beta) \quad \text{if} \quad \phi(x, \alpha, \beta) \geq 0,$$
$$\phi(x, \alpha, \beta) \leq \alpha\Phi(x, \alpha, \beta) \quad \text{if} \quad \phi(x, \alpha, \beta) \leq 0.$$

Inequalities (3.6.8) and (3.6.9) demonstrate that $\Phi(x, \alpha, \beta)$ and $\phi(x, \alpha, \beta)$ have the same signs and that if $\phi(x, \alpha, \beta) \geq 0$,

$$\frac{\phi(x, \alpha, \beta)}{\beta} \leq \Phi(x, \alpha, \beta) \leq \frac{\phi(x, \alpha, \beta)}{\alpha};$$

and if $\phi(x, \alpha, \beta) \leq 0$,

$$\frac{\phi(x, \alpha, \beta)}{\alpha} \leq \Phi(x, \alpha, \beta) \leq \frac{\phi(x, \alpha, \beta)}{\beta}.$$

Since f is bounded, there exists a number M such that $|f(x)| \leq M$ for all x. Thus,

$$|f(x + h) - f(x)| \leq 2M.$$

Therefore,

$$\left| \Phi(x, \alpha, \beta) - \frac{\phi(x, \alpha, \beta)}{\alpha} \right| \leq |\phi(x, \alpha, \beta)| \frac{\beta - \alpha}{\alpha\beta} \leq 2M \frac{\beta - \alpha}{\alpha\beta}.$$

Consequently, by (3.6.6)

(3.6.10)
$$\left| \Phi(x, a_{k-1}, a_k) - \frac{\phi(x, a_{k-1}, a_k)}{a_{k-1}} \right| \leq 2M \frac{b - a}{na^2},$$

for $k = 1, 2, \ldots, n$. Let

(3.6.11)
$$G(x, a, b, n) = \max_{k \leq n} \frac{\phi(x, a_{k-1}, a_k)}{a_{k-1}}.$$

By (3.6.9), we have

$$|\Phi(x, a, b) - G(x, a, b, n)| \leq \frac{2M(b - a)}{na^2}.$$

Hence,

(3.6.12)
$$\Phi(x, a, b) = \lim_{n \to \infty} G(x, a, b, n).$$

Moreover, the sequence on the right is uniformly convergent. We conclude that

(3.6.13)
$$D^+f(x) = \lim_{j \to \infty} \lim_{k \to \infty} \Phi\left(x, \frac{1}{k}, \frac{1}{j}\right).$$

If f is a bounded function of class $\alpha > 0$, by Corollary 3.6.3 the functions

$$\frac{\phi(x, a_{k-1}, a_k)}{a_{k-1}}$$

are of class $\leq \alpha$; therefore, so is G. The sequence in (3.6.12) converges uniformly; thus, the limit function Φ is of class $\leq \alpha$. Hence, D^+f is of class $\leq \alpha + 2$.

Now suppose f is measurable and unbounded. If p is a positive integer, define

$$f_p(x) = \begin{cases} f(x), & |f(x)| \leq p \\ p, & f(x) > p \\ -p, & f(x) < -p. \end{cases}$$

The each f_p is bounded and measurable. It is easy to see that

$$D^+f(x) = \lim_{p \to \infty} D_p^+ f(x)$$

from which it follows that D^+f is measurable. $\qquad\qquad\qquad\square$

§3.7. Dini Derivatives and Convex Functions

We present in this section a collection of results on properties of convex functions using Dini derivatives.

Definition 3.7.1. Let $I \subset \mathbf{R}^1$ be an open interval. A function $f: I \to \mathbf{R}^1$ is *convex* if for each $x, y \in I$ and each $\lambda \in [0, 1]$, we have

$$f[\lambda x + (1 - \lambda)y] \leq \lambda f(x) + (1 - \lambda)f(y).$$

Geometrically, this means that if we look at the graph of f in $\mathbf{R}^2$, then any chord connecting $(x, f(x))$ and $(y, f(y))$ will lie above the graph of f.

Lemma 3.7.2 [R4, p. 113]. *If f is convex on the open interval I and if $x, y, x', y' \in I$ with $x \leq x' < y'$ and $x < y \leq y'$, then*

$$\frac{f(y) - f(x)}{y - x} \leq \frac{f(y') - f(x')}{y' - x'}.$$

Proof. Let $\lambda = (y' - x')/(y' - x)$. Then $1 - \lambda = (x' - x)/(y' - x)$ and $\lambda x + (1 - \lambda)y' = x'$. Applying the definition, we get

$$f(x') \leq \frac{y' - x'}{y' - x} f(x) + \frac{x' - x}{y' - x} f(y'),$$

which is equivalent to

$$0 \leq \frac{y' - x'}{y - x}[f(x) - f(x')] + \frac{x' - x}{y' - x}[f(y') - f(x')]$$

which is, in turn, equivalent to

$$\frac{f(x') - f(x)}{x' - x} \leq \frac{f(y') - f(x')}{y' - x'}.$$

Now let $\lambda = (y' - y)/(y' - x)$ to yield

$$\frac{f(y) - f(x)}{y - x} \leq \frac{f(x') - f(x)}{x' - x}. \qquad \square$$

This result also has a geometric interpretation: The slope of the chord connecting the points $(x', f(x'))$ and $(y', f(y'))$ is smaller than (or equal to) the slope of the chord connecting the points $(x, f(x))$ and $(y, f(y))$.

Theorem 3.7.3 [R4, p. 113]. *If f is convex on the open interval I, then f is Lipschitz continuous on each closed subinterval of I.*

Proof. Let J be any closed interval contained in I. Choose points $a < b$ in I which are to the left of J and points $c < d$ in I which are to the right of J. Then by the preceding lemma we have

$$\frac{f(b) - f(a)}{b - a} \leq \frac{f(y) - f(x)}{y - x} \leq \frac{f(d) - f(c)}{d - c}$$

for all $x, y \in J$. Thus, there exists a constant M such that

$$|f(y) - f(x)| \leq M|x - y| \quad \text{for all } x, y \in J;$$

that is, f is Lipschitz continuous on J. $\qquad \square$

Theorem 3.7.4 [R4, p. 113]. *If f is convex on the open interval (a, b), the right and left derivatives of f exist at each point of (a, b) and are equal except on a countable set. The left and right derivatives are increasing functions, and at each point, the left derivative is less than or equal to the right derivative.*

Proof. Let $x_0 \in (a, b)$. By Lemma 3.7.2,

$$\frac{f(x) - f(x_0)}{x - x_0}$$

is an increasing function in x, so the limits as x approaches x_0 from the right and left exist and are finite. Hence, f has a right and left derivative at each point and the left derivative is less than or equal to the right derivative. If

$x_0 < y_0$, $x < y_0$, and $x_0 < y$, then

$$\frac{f(x) - f(x_0)}{x - x_0} \leq \frac{f(y) - f(y_0)}{y - y_0},$$

so that either derivative at x_0 is less than or equal either derivative at y_0. Thus, each derivative is monotone and they are equal to a point where one of them is continuous (Theorem 3.4.6). Since a monotone function can have at most a countable set of discontinuities, the right and left derivatives must be equal except on a countable set.   □

EXERCISES

1. Show that if f is continuous on $[a, b]$ and if any one of D^+f, D_+f, D^-f, or D_-f is non-negative on $[a, b]$, then $f(a) \leq f(b)$.

2. Show that if f has a relative maximum at 0, then $D^+f(0) \leq 0$ and $D_-f(0) \geq 0$.

3. Let f be increasing and finite in $[a, b]$. Then for almost all x, f has a finite derivative $f'(x)$. Let $g(x) = f'(x)$ where $f'(x)$ is defined and finite and $g(x) = 0$ elsewhere. Then g is integrable and

$$f(x) = \int_a^x g(x)\,dx + h(x),$$

where h is an increasing function with $h'(x) = 0$ a.e.

4. If the right derivative f'_+ of a function f is finite except on a set of points which is at most denumerable, then f is Borel measurable and of Baire class ≤ 2. [**Hint**: Let D be the set (at most denumerable) where f'_+ is infinite. Let a be any real number and

$$E = \{x : f(x) \geq a\}.$$

Show that the points $x \notin E$ which are accumulation points from the right of E (i.e., points which are the limit of a decreasing sequence of points in E) form at most a denumerable set $D_1 \subset D$. Similarly, the points which are accumulation points of E from the left form at most a denumerable set D_2. Since the union of the sets E, D_1, and D_2 is closed, E is a G_δ set. Therefore, f is a function of class ≤ 2.]

5. If the right derivative f'_+ is finite a.e., f is a (Lebesgue) measurable function. [**Hint**: The proof is similar to the preceding one. Show that the union of the set

$$E = \{x : f(x) \geq a\}$$

and a set of measure zero is closed, from which it follows that f is measurable.]

6. Prove the following consequences of the preceding results:

   (a) A function which is not Borel measurable has one infinite Dini derivative on an nondenumerable set of points.
   (b) A function which is not (Lebesgue) measurable has one infinite Dini derivative on a set of positive measure or on a set which is not (Lebesgue) measurable.

7. **Definition** [S1, p. 108]. Let f and u be two finite functions and let x_0 be a point such that u is not identically constant in any interval containing x_0. The *upper derivate of f with respect to u at x_0* is denoted $\overline{D}_u f(x_0)$ and is the upper limit of the ratio

$$\frac{f(x) - f(x_0)}{u(x) - u(x_0)}$$

as $x \to x_0$ by values other than those for which

$$f(x) - f(x_0) = u(x) - u(x_0) = 0.$$

The *lower derivate of f with respect to u at x_0* is denoted $\underline{D}_u f(x_0)$ and is defined as the lower limit of such ratios. Similarly, we can define the four Dini derivatives of f with respect to u: $D_u^+ f(x)$, $D_{u^+} f(x)$, $D_u^- f(x)$, $D_{u^-} f(x)$.

Show that if f and u are two finite functions, the set of points t at which the derivative $u'(t) > 0$ (finite or infinite) exists and at which $D_u^+ f(t) < D_{u^-} f(t)$ is at most denumerable. [**Hint:** use Theorem 3.4.8.]

8. Show for every increasing function f,

$$D^+ f(x) < +\infty$$

and

$$D^+ f(x) < D_- f(x)$$

almost everywhere.

9. Show that

$$D^+ f(x) = D_+ f(x) < +\infty$$

if and only if

$$\lim_{t \to 0, t > 0} \frac{f(x+t) - f(x)}{t}$$

exists. Prove the analogous result for the left derivates.

10. Let f be an increasing function. Define

$$F(x) = f(x+) = \lim_{h \to 0, h > 0} f(x + h).$$

Show F is increasing, right continuous, and upper semicontinuous.

11. Let f and F be as defined in Exercise 10. If x is a point of continuity of f, show

$$D^+ f(x) \le D^+ F(x)$$

and

$$D_- F(x) \le D_- f(x).$$

Thus, $D^+ f(x) < \infty$ and $D^+ f(x) \le D_- f(x)$ hold a.e. if the same inequalities hold a.e. for F.

12. For f upper semicontinuous, right continuous, and increasing, show

$$\{x: D^+f(x) = +\infty\}$$

is a set of measure zero.

13. Let f be as given in Exercise 12. Let p and q be any two rational numbers such that $0 < p < q$. Let

$$E_{pq} = \{x: D^+f(x) > q > p > D_-f(x)\}.$$

For any interval (a, b), show $E_{pq} \cap (a, b)$ is contained in the union of a sequence of open intervals of length less than or equal $(p/q)(b - a)$.

14. Let f be as given in Exercise 12. Show that

$$\{x: D^+f(x) > D^-f(x)\}$$

is a set of measure zero. (The results of Exercises 12–14 give another proof that an increasing function is differentiable a.e.)

15. Construct an increasing jump function whose derivative does not exist everywhere off the set of its discontinuities. [**Hint:** let the graph of the function oscillate at 0 between the lines $y = x$ and $y = 2x$.]

16. [H2, p. 402] Let G be a perfect nondense set of measure zero in the interval $[a, b]$. Let u_n be the length of the contiguous interval $[a_n, b_n]$. Suppose that $u_1, u_2, \ldots, u_n,$ $\ldots$ are in descending order of magnitude. Let $h_1, h_2, \ldots, h_n, \ldots$ be positive numbers such that $\sum h_n u_n$ is convergent and let

$$y_n = \frac{2}{\pi} h_n u_n \sin^{-1}\left(\frac{x - a_n}{u_n}\right)^{1/2}, \quad x \in [a_n, b_n].$$

Let

$$f(x) = (a \sum x) h_n u_n + \omega y_n(x),$$

where ω is zero if $x \in G$ and is 1 if x is in the interior of an interval contiguous to G; $(a \sum x) h_n u_n$ denotes the sum of those parts of $\sum h_n u_n$ that are in the interval $[a, x]$. Show that f is continuous; it has a finite derivative at every point not in G; and it has an infinite derivative at all points of G.

17. [H2, p. 402] With the same notation as in Exercise 16, let

$$f(x) = \begin{cases} 0, & x \in G \\ u_n^2 y\left(\dfrac{x - a_n}{u_n}\right), & x \in [a_n, b_n], \end{cases}$$

where

$$y(x) = (1 - x)^2 x^{1/2} \sin^2\left(\frac{1}{x}\right).$$

Show f is continuous and it has a finite and continuous derivative at all points not in G. At the endpoints of all contiguous intervals, $D^+f(x) = +\infty$. The other three Dini derivatives are all zero at all points of G.

18. For any subset E of the real line and any function f defined on the real line, let f_E be the function defined on the interval $(0, b)$, where $b = \sup\{z: z \in E\}$, by

$$f_E(x) = \sup\{f(y): 0 < y \le x\}.$$

Use the fact that a monotone function is differentiable a.e. to prove the following: For any f defined on $\mathbf{R}^1$, let

$$E = \{x: x > 0 \text{ and } f(x) - f(y) > 0 \ \forall y \in (0, x)\}.$$

Show

$$f_E'(x) = \lim_{h \to 0} \left\{ \frac{f(x+h) - f(x)}{h}: x, x + h \in E \right\}$$

exists a.e. on E.

19. For f_E' as given in Exercise 18, show

$$f_E'(x) = D^+ f(x)$$

a.e. on E.

20. For f_E' as given in Exercise 18, show

$$f_E'(x) = D_- f(x)$$

a.e. on E.

21. Suppose f is a continuous function defined on the interval $[a, b]$ and

$$\frac{f(b) - f(a)}{b - a} < c.$$

Show there are uncountably many $x \in [a, b]$ where $D^+ f(x) < c$. [**Hint:** Let

$$g(x) = f(x) - f(a) - \left(\frac{x - a}{b - a} \right) k,$$

where $k > f(b) - f(a)$. Then $g(a) = 0$ and $g(b) < 0$. Let s be such that $0 = g(a) > s > g(b)$ and $x_0 = \sup\{x \in [a, b]: g(x) \ge s\}$. Show $D^+ g(x_0) \le 0$ and thus $D^+ f(x_0) < c$. But since s was any arbitrary point in the interval $(g(b), g(a))$, the desired result is obtained.]

22. Use Problem 21 to show that if f is continuous on $[a, b]$ and if any one of the Dini derivatives is non-negative except for perhaps a countable number of points in $[a, b]$, then f is increasing (cf. Theorem 3.4.4).

23. Show that continuity is critical in the previous problem by constructing a discontinuous function f, which is not increasing, such that $D_+ f(x) \ge 0$ for every x.

24. This problem illustrates that continuity is necessary in Corollary 3.5.8. We will use the ternary scale to represent the real numbers in the interval $(0, 1)$ (see Example 1.3.3); when there is more than one ternary representation for a number, we will use the one that terminates. If $x = {}_{.3}a_1 a_2 \ldots$, then $f(x) = {}_{.2}b_1 b_2 \ldots$, where $b_n = 1$ if $a_n = 2$, and $b_n = 0$ otherwise. (**Note:** This is **not** the Cantor ternary function which is constant on the intervals removed in the construction of the Cantor ternary set.) Since ternary representations ending in repeated 2's excluded,

the ternary representation of every $x \in (0, 1)$ contains an infinite sequence of either 0's or 1's. Let one of these digits (either 0 or 1) occur at the rth place in the ternary representation of x. Let x' differ from x only by having 2 as its rth ternary digit. Show that

$$\frac{f(x') - f(x)}{x' - x} \geq \frac{3^r}{2^r}.$$

Since r is arbitrarily large, $D^+f(x) = \infty$. Thus, if f is not required to be continuous, then $D^+f(x)$ may be $+\infty$ at every x.

25. Let f and g be defined on $[0, 1]$. Let f have jumps of $m_n = 2^{-n}$ at $r_n = 3^{-n}$, $n = 1$, $2, \ldots$; let g have jumps $p_n = 3^{-n}$ at $q_n = 2^{-n}$, $n = 1, 2, \ldots$. Show that f and g are continuous from the right at 0, but $Df_+(0) = \infty$, whereas $Dg_+(0) = 0$.

26. Let $f: [a, b] \to \mathbf{R}^1$ be continuous on $[a, b]$ and differentiable on (a, b). If $\lim_{x \to a+} f'(x) = A$ (A may be finite or infinite), then $f'_+(a)$ exists and is equal to A.

CHAPTER 4

Approximate Derivatives

Throughout this chapter we suppose that f is a function which is measurable and finite. We begin by generalizing the concept of the density of a set (see Chapter 2) in a manner similar to the generalization of derivatives in the last chapter. We then introduce the concept of approximate derivatives and prove that the approximate derivatives of a measurable function are measurable. Analogous to the Denjoy–Saks–Young Theorem for Dini derivatives, we establish a complete characterization of the approximate derivatives of a function. Baire category results are given in Section 4.4. We then conclude the chapter with a collection of results on other properties of approximate derivatives, notably the Darboux property and the mean value property. Thus, this chapter is a generalization of the standard results on derivatives.

§4.1. Definitions

Definition 4.1.1 [BH2]. Let E be a measurable set. The *upper right-hand density* of E at the point x (not necessarily in E), denoted $d^+(E, x)$, and the *lower right-hand density* of E at x, denoted $d_+(E, x)$, are defined as the upper and lower limits, respectively, of

$$\frac{m[E \cap (x, x + h)]}{h}$$

as $h \to 0$. Similarly, the *upper left-hand density* of E at x, denoted $d^-(E, x)$, and the *lower left-hand density* of E at x, denoted $d_-(E, x)$, are defined as the upper and lower limits, respectively, of

$$\frac{m[E \cap (x - h, x)]}{h}$$

as $h \to 0$. If all four limits are equal, their common value is the density of E at x, denoted, $d(E, x)$. As in Chapter 2, we can extend these concepts to non-measurable sets; we simply replace measure m by outer measure m^* in the above definitions.

Definition 4.1.2 [BH2]. For a fixed x in the domain of the function f and a given $\lambda < 1$, consider the set of all numbers a such that the set of points ξ for which

$$\frac{f(\xi) - f(x)}{\xi - x} \geq a, \quad \xi > x,$$

has at x upper right-hand density $\leq \lambda$. The *upper right λ-derivate*, $AD^+(f, x, \lambda)$, is the lower bound of the set of all such a. As λ decreases, $AD^+(f, x, \lambda)$ increases. The *upper right approximate derivate*, $AD^+f(x)$, is

$$\lim_{\lambda \to 0} AD^+(f, x, \lambda).$$

There are corresponding definitions for the lower right approximate derivative AD_+ and the left derivates AD^- and AD_-. When the upper and lower right approximate derivates are equal, we call their common value the *right-hand approximate derivate*. When all four approximate derivates are equal, we call their common value the *approximate derivative*.

§4.2. Measurability of Approximate Derivatives

To prove that λ-derivates and approximate derivates are measurable (Theorem 4.2.4), we require some preliminary results.

Lemma 4.2.1 [BH2]. *Let*

$$M(x, x + h) = m[\{\xi: \xi \in (x, x + h) \text{ and } f(\xi) \geq f(x)\}].$$

Then, for fixed h, $M(x, x + h)$ is a measurable function of x.

Proof. Let $\lambda > 0$ and

$$E = \{x: M(x, x + h) < \lambda h\}.$$

Let $\phi(x, x + h, \lambda)$ be the upper bound (which is attained) of all the numbers c for which

$$m\{x': x' \in (x, x + h) \text{ and } f(x') \geq c\} \geq \lambda h.$$

Then $E = \{x: f(x) > \phi(x, x + h, \lambda)\}$. To prove this, suppose $x \in E$. Then $f(x) > c$ for every c in the definition of $\phi(x, x + h, \lambda)$. For if not, there exists some c' such that

$$m\{x': x' \in (x, x + h) \text{ and } f(x') \geq c'\} \geq \lambda h,$$

but $f(x) \leq c'$. Then if $x' \in (x, x + h)$ and $f(x') \geq c'$, we have $f(x') \geq c' \geq f(x)$. Thus, $x' \in \{\xi: \xi \in (x, x + h) \text{ and } f(\xi) \geq f(x)\}$. Hence, we would have

$$\lambda h \leq m\{x': x' \in (x, x + h) \text{ and } f(x') \geq c'\}$$
$$\leq m\{\xi: \xi \in (x, x + h) \text{ and } f(\xi) \geq f(x)\} < \lambda h,$$

which is obviously a contradiction. Since $f(x) > c$ for all c in the definition of $\phi(x, x + h, \lambda)$, we have $f(x) > \phi(x, x + h, \lambda)$ for every $x \in E$. Conversely, if $f(x) > \phi(x, x + h, \lambda)$, then $x \in E$ by similar reasoning.

Now if we can show, for fixed h and λ, $\phi(x, x + h, \lambda)$ is a measurable function of x, then the set E will be measurable and we will have the desired result. We will show that $\phi(x, x + h, \lambda)$ is upper semicontinuous. Suppose $\phi(x_0, x_0 + h, \lambda) = c_0$. We need to show that

$$\overline{\lim_{x \to x_0}} \; \phi(x, x + h, \lambda) \leq \phi(x_0, x_0 + h, \lambda) = c_0.$$

Recall that $\phi(x_0, x_0 + h, \lambda)$ is the upper bound of all c such that

$$m\{x': x' \in (x, x + h) \text{ and } f(x') \geq c\} \geq \lambda h.$$

Thus, for $\varepsilon > 0$, there exists $\delta > 0$ such that

$$m[\{x': x' \in (x_0, x_0 + h) \text{ and } f(x') \geq c_0 + \varepsilon\}] = \lambda h - \delta < \lambda h.$$

Now let $x \in (x_0 - \delta, x_0 + \delta)$ and consider the interval $(x, x + h)$. Suppose $x < x_0$. Then

$$m[\{x': x' \in (x, x + h) \text{ and } f(x') \geq c_0 + \varepsilon\}]$$
$$\leq m[\{x': x' \in (x, x_0) \text{ and } f(x') \geq c_0 + \varepsilon\}]$$
$$+ m[\{x': x' \in (x_0, x + h) \text{ and } f(x') \geq c_0 + \varepsilon\}]$$
$$\leq \delta + m[\{x': x' \in (x_0, x_0 + h) \text{ and } f(x') \geq c_0 + \varepsilon\}]$$
$$= \delta + \lambda h - \delta = \lambda h.$$

The case $x > x_0$ is similar. Therefore, for $x \in (x_0 - \delta, x_0 + \delta)$, $\phi(x, x + h, \lambda) \leq c_0 + \varepsilon$. Thus,

$$\overline{\lim_{x \to x_0}} \; \phi(x, x + h, \lambda) \leq c_0 + \varepsilon.$$

Since ε is arbitrary, we have proved the upper semicontinuity of $\phi(x, x + h, \lambda)$ and hence its measurability.   □

Lemma 4.2.2 [BH2]. *Fix x and h and let*

$$E = \{\xi: \xi \in (x, x + h) \text{ and } f(\xi) \geq f(x)\}.$$

Let

$$\alpha(x) = d_+(E, x) = \varliminf_{h \to 0} \frac{M(x, x + h)}{h}$$

and

$$\beta(x) = d^+(E, x) = \varlimsup_{h \to 0} \frac{M(x, x + h)}{h}.$$

The functions $\alpha(x)$ and $\beta(x)$ are measurable.

Proof. Let $\underline{M}(x, x + h)$ be the lower bound of $\{M(x, x + j)/j: 0 < j < h\}$. Since $M(x, x + j)/j$ is a continuous function of j in this range, the lower bound will be the same if only rational values of j are considered. Then $\underline{M}(x, x + h)$ is the lower bound of a sequence of measurable functions and hence is measurable. As h decreases, $\underline{M}(x, x + h)$ increases. Thus,

$$\alpha(x) = \lim_{h \to 0} \underline{M}(x, x + h)$$

is measurable. A similar proof will hold for $\beta(x)$.   □

Corollary 4.2.3 [BH2]. *Let $\alpha_a(x)$ and $\beta_a(x)$ be the lower and upper densities, respectively, at x of the set*

$$\{\xi: \xi > x \text{ and } f(\xi) - f(x) \geq a(\xi - x)\}.$$

Then $\alpha_a(x)$ and $\beta_a(x)$ are measurable.

Proof. Apply the proof of Lemma 4.2.2 to the function $f(x) - ax$.   □

Theorem 4.2.4 [BH2]. *The λ-derivates and the approximate derivates of a measurable function are measurable.*

Proof. Let

$$E = \{x: AD^+(f, x, \lambda) \leq a\}.$$

Let a_n be a decreasing sequence which converges to a. Let E_n be the set of x for which the set

$$\{\xi: \xi > x \text{ and } f(\xi) - f(x) \geq a_n(\xi - x)\}$$

has upper right-hand density at x less than or equal to λ. Then the sets E_n form a decreasing sequence with limit E. But $E_n = \{x: \beta_{a_n}(x) \leq \lambda\}$ and hence is measurable by Corollary 4.2.3. Therefore, E is measurable and $AD^+(f, x, \lambda)$ is measurable. The proof for the other λ-derivates is similar. Since the approximate derivates are the limits of the λ-derivates, they too are measurable.   □

§4.3. Analogue of the Denjoy–Saks–Young Theorem

Lemma 4.3.1 [BH2]. *The set of points x such that*

$$f(x') - f(x) < a(x' - x)$$

for all $x' \in (x, x + j)$, with the possible exception of a set of measure less than $(\lambda + \varepsilon)j$, where $j \in (0, h]$, is measurable.

Proof. Define $f_1(x) = f(x) - ax$. Let A_j be the set of points x such that

$$M(x, x + j) = m\{x': x' \in (x, x + j) \text{ and } f_1(x') \geq f_1(x)\} < (\lambda + \varepsilon)j.$$

Thus, for each $x \in A_j$, $f_1(x') < f_1(x)$ for all $x' \in (x, x + j)$ except for a set of measure less then $(\lambda + \varepsilon)j$; that is, $f(x') - f(x) < a(x' - x)$ for all $x' \in (x, x + j)$ except for a set of measure less than $(\lambda + \varepsilon)j$. As in the proof of Lemma 4.2.2, we consider only rational values of j. Thus, the set described in the statement of this lemma is the intersection of the denumerable sequence of sets $\{A_j\}$. The proof of Lemma 4.2.1 shows that each set A_j is measurable; hence, the intersection of this sequence of sets is measurable. □

Theorem 4.3.2 [BH2]. *Suppose that at points x of a measurable set E of finite measure, for some $\lambda < 1$,*

$$AD^+(f, x, \lambda) < +\infty.$$

Then f has a finite approximate derivative a.e. in E.

Proof. Let $0 < \varepsilon < 1 - \lambda$. Because f and $AD^+(f, x, \lambda)$ are measurable, we can use Lusin's theorem to choose a and M (sufficiently large) so that

$$AD^+(f, x, \lambda) < a,$$
$$-M < f(x) < M$$

in a set of measure greater than $mE - \varepsilon$. Choose h small enough so that we can find a set E_1 of measure greater than $mE - 2\varepsilon$ such that, for $x \in E_1$,

$$f(x') - f(x) < a(x' - x)$$

for all $x' \in (x, x + j)$ with the possible exception of a set of measure less than $(\lambda + \varepsilon)j$, $j \in (0, h]$. By Lemma 4.3.1, E_1 is measurable. We can find a subset $E_2 \subset E_1$ and numbers γ (small) and $k \in (\lambda + \varepsilon, 1)$ such that if $x \in E_2$,

(a) $m[E_1 \cap (x - \delta, x)] > k\delta$ if $\delta \leq \gamma$,
(b) f is approximately continuous at x,
(c) $mE_2 > mE - 3\varepsilon$.

Take any interval (b, c) whose length is less than the $\min(h, \gamma)$ and which contains points of E_2. Let

$$f_1(x) = f(x) - ax$$

and define $\phi(x)$ to be the lower bound of the set

$$\{f_1(\xi): \xi \in E_2 \cap [b, x]\}.$$

Clearly, $f_1(x) \geq \phi(x)$ for $x \in E_2 \cap (b, c)$. We will prove that, at all these points, $f_1(x) = \phi(x)$. Suppose, on the contrary, that for some $x_1 \in E_2$ and some $\eta > 0$,

$$f_1(x_1) = \phi(x_1) + \eta.$$

We shall construct a sequence of points $\xi_n \in E_1$ such that ξ_n converges to x_1 from the left and

$$f_1(\xi_n) < f_1(x_1) - \tfrac{1}{2}\eta.$$

Then by definition of E_1, the set of points in (ξ_n, x_1) such that

$$f_1(x) > f_1(x_1) - \tfrac{1}{2}\eta$$

has measure less than $(\lambda + \varepsilon)(x_1 - \xi_n)$. Since $\lambda + \varepsilon < 1$ and ξ_n may be arbitrarily near x_1, this inequality contradicts the approximate continuity of $f_1(x)$ at x_1. Thus, we conclude that $f_1(x_1) = \phi(x_1)$.

Now we construct the sequence ξ_n. Let ξ_0 be any point in $E_2 \cap (b, x_1)$ for which

$$f_1(\xi_0) < f_1(x_1) - \tfrac{1}{2}\eta.$$

The set of points $x \in (\xi_0, x_1)$ for which $f_1(x) > f_1(\xi_0)$ has measure less than $(\lambda + \varepsilon)(x_1 - \xi_0)$ and the set of points $x \in (\xi_0, x_1) \cap E_1$ has measure greater than $k(x_1 - \xi_0)$ [condition (a) above]. Hence, there is a point

$$\xi_1 \in E_1 \cap (\xi_0 + \{k - \lambda - \varepsilon\}\{x_1 - \xi_0\}, x_1)$$

for which

$$f_1(\xi_1) \leq f_1(\xi_0) < f_1(x_1) - \tfrac{1}{2}\eta.$$

Repeating this argument for the interval (ξ_1, x_1), we obtain a point ξ_2 lying in an interval of length $(1 - k + \lambda + \varepsilon)^2(x_1 - \xi_0)$ to the left of x_1, and so on. Since $k - \lambda - \varepsilon > 0$, we have $\xi_n \to x_1$ as $n \to \infty$. Thus, ξ_n is the required sequence.

Now let the function $f(x, E_2) = f(x)$ if $x \in E_2$; $f(x, E_2)$ is undefined elsewhere on E. If we cover E by intervals of length less than $\min(h, \gamma)$, we have proved that in any such interval (b, c) which contains points of E_2, $f_1(x, E_2)$ is a decreasing function in the set E_2. Hence, at almost all points of E_2, $f_1(x, E_2)$ has a finite derivative. Therefore, $f(x, E_2)$ has a finite derivative $f'(x, E_2)$ a.e. in E_2.

At any point of density of E_2 for which $f'(x, E_2)$ exists, f has an approximate derivative equal to $f'(x, E_2)$. Thus, we have shown that f has a finite approximate derivative at points of a set whose measure is greater than $mE - 3\varepsilon$. Since ε was arbitrary, a finite approximate derivative exists a.e. in E.

$\square$

Corollary 4.3.3 [BH2]. *The set in which $AD^+ f(x) = -\infty$ has measure zero.*

Corollary 4.3.4 [BH2]. *Theorem 4.3.2 is true if λ depends on x.*

Proof. If $AD^+(f, x, \lambda) < +\infty$ in E and $\varepsilon > 0$, we can fix λ_0 such that $AD^+(f, x, \lambda_0) < +\infty$ in a set of measure greater than $mE - \varepsilon$. $\qquad\square$

Corollary 4.3.5 [BH2]. *Except on a set of measure zero,*

$$AD^+ f(x) = AD^+(f, x, \lambda)$$

for all $\lambda \in (0, 1)$.

Proof. This result follows from the previous corollary. $\qquad\square$

The next result shows the connection between approximate derivatives and Dini derivatives.

Theorem 4.3.6 [BH2]. *If $D^+ f(x)$ is finite in E, then a.e. in E,*

$$D^+ f(x) = ADf(x) = D_- f(x).$$

Proof. Recall

$$D^+ f(x) = \overline{\lim_{h \to 0^+}} \frac{f(x + h) - f(x)}{h}.$$

Thus, if $D^+ f(x) < +\infty$ in E, then $AD^+ f(x) < +\infty$ in E and, by Theorem 4.3.2, a finite approximate derivative $ADf(x)$ exists a.e. in E.

Let

$$S = \{x \in E : D^+ f(x) > ADf(x)\}.$$

Suppose $mS > 0$. By Lusin's theorem, there exist positive numbers B, b, and a set E_0, whose measure is greater than zero, in which

$$D^+ f(x) < B,$$

$$ADf(x) > -B,$$

$$D^+ f(x) - ADf(x) > b.$$

Let $h_1 > 0$ and let E_1 be the set of points $x \in E_0$ such that

(4.3.7) $$f(x') - f(x) < B(x' - x)$$

for all $x' \in (x, x + h_1]$. Then E_1 is measurable (by Theorem 3.6.5) and we may take h_1 small enough that $mE_1 > 0$.

Let x_0 be a point of density of E_1 and let ε and η be small positive numbers. We can find $h_0 \leq h_1$ such that the set E_2 of points x in

$E_1 \cap (x_0, x_0 + h)$ for which

(4.3.8) $f(x) - f(x_0) < \{ADf(x_0) + \eta\}(x - x_0)$

has measure at least $(1 - \varepsilon)h$ whenever $h \le h_0$. But there are points x', arbitrarily near x_0, for which

(4.3.9) $f(x') - f(x_0) > \{D^+f(x_0) - \eta\}(x' - x_0).$

If $x' \in (x_0, x_0 + h)$, there is a point $x \in E_2$ such that $x' - x < \varepsilon(x' - x_0)$. Combining (4.3.8) and (4.3.9), we obtain

$$f(x') - f(x) > \{D^+f(x_0) - ADf(x_0) - 2\eta\}(x' - x_0) + \{ADf(x_0) + \eta\}(x' - x)$$

$$> \left(\frac{b - 2\eta}{\varepsilon} - B\right)(x' - x).$$

For sufficiently small ε, this contradicts (4.3.7). Hence, $mS = 0$. By a similar argument, we can show that the set in which $D_- f(x)$ is less than $ADf(x)$ has measure zero.   □

We now present the analogue of the Denjoy–Saks–Young Theorem for approximate derivatives.

Theorem 4.3.10 [BH2]. *If f is a finite measurable function, then a.e. either*

(a) *a finite approximate derivative exists, or*
(b) $AD^+f(x) = AD^-f(x) = +\infty, AD_+f(x) = AD_-f(x) = -\infty.$

Except on a set of measure zero, the points where a finite approximate derivative exists are of three types:

(c) *A finite ordinary derivative exists.*
(d) $D_+f(x) = D^-f(x) = ADf(x), D^+f(x) = +\infty, D_-f(x) = -\infty,$ *or* $D^+f(x) = D_-f(x) = ADf(x), D_+f(x) = -\infty, D^-f(x) = +\infty.$
(e) $D^+f(x) = D^-f(x) = +\infty, D_+f(x) = D_-f(x) = -\infty.$

Proof. Theorem 4.3.2 and Corollaries 4.3.3–4.3.5 give (a) and (b). In fact, at points where (b) holds, we have every upper λ-derivate is $+\infty$ and every lower λ-derivate is $-\infty$. Theorem 4.3.6 gives the remainder of the theorem.

§4.4. Category Results for Approximate Derivatives

Our next major result is that approximate derivatives are functions of Baire class 1 (Theorem 4.4.18). We begin this section with a discussion of *interval functions*.

Definition 4.4.1 [B2, p. 103]. Let $\{I\}$ denote the collection of all nondegenerate closed intervals contained in some closed interval I_0. To say that $I \to x$

will mean that $x \in I$ and $mI \to 0$. Let ϕ be a real-valued function defined on $\{I\}$. We say that ϕ is a *convergent interval function* if, for each $x_0 \in I_0$,

$$\lim_{I \to x_0} \phi(I)$$

exists.

EXAMPLE 4.4.2 [B2, p. 104]. If f is a real function and ϕ is the interval function defined by

$$\phi([a, b]) = f(b) - f(a),$$

then ϕ is convergent if and only if f is continuous. If

$$\phi([a, b]) = \frac{f(b) - f(a)}{b - a},$$

then ϕ is convergent if and only if f is differentiable.

Theorem 4.4.3 [B2, p. 104]. *Let f be a real-valued function defined on I_0. Then f is a Darboux function of Baire class 1 if and only if for each $I \in \{I\}$ there exists a point x_I in the interior of I such that if $x \in I_0$, then $f(x_I) \to f(x)$ whenever $I \to x$.*

Proof. We will first show that f is in Baire class 1 whenever the condition is met. If f is not in Baire class 1, then there exists a perfect set P such that $f|_P$ has no point of continuity. Thus, the oscillation of $f|_P$ at x, $\omega(f|_P, x)$, is greater than zero for all $x \in P$. For each positive integer n, let

$$E_n = \left\{ x: \omega(f|_P, x) \geq \frac{1}{n} \right\}.$$

There exists an n such that E_n is somewhere dense in P. Since E_n is closed, E_n contains a relative interval of P. Thus, there exists a nonempty perfect set $Q \subset P$ such that $\omega(f|_Q, x) \geq 1/n$ for all $x \in Q$.

For each positive integer k, let

$$Q_k = \left\{ x \in Q: |f(x_I) - f(x)| < \frac{1}{10n} \text{ whenever } mI < \frac{1}{k} \text{ and } x \in I^0 \right\}.$$

Then $Q = \bigcup Q_k$; thus, there exists an open interval J such that Q_k is dense in $Q \cap J$. We can shorten J if necessary so that $mJ < 1/k$.

Let $x_1 \in J \cap Q_k$. Then

(4.4.4) $$|f(x_1) - f(x_J)| < \frac{1}{10n}.$$

Since $\omega(f|_Q, x) \geq 1/n$ for all $x \in Q$, there exists an $x_2 \in J \cap Q$ such that

(4.4.5) $$|f(x_2) - f(x_1)| \geq \frac{1}{2n}.$$

Using the hypothesis of the theorem, choose an open interval $J_1 \subset J$ such that $x_2 \in J_1$ and

(4.4.6)
$$|f(x_{J_1}) - f(x_2)| < \frac{1}{10n}.$$

Finally, choose $x_3 \in J_1 \cap Q_k$. We can choose such a point because $x_2 \in Q$ and Q_k is dense in Q. Then since $mJ_1 < 1/k$,

$$|f(x_{J_1}) - f(x_3)| < \frac{1}{10n}.$$

Combining with (4.4.6), we conclude

(4.4.7)
$$|f(x_2) - f(x_3)| < \frac{2}{10n}.$$

From (4.4.4) and the corresponding inequality for x_3 (which holds because $x_3 \in J \cap Q_k$), we obtain

(4.4.8)
$$|f(x_1) - f(x_3)| < \frac{2}{10n}.$$

Then from (4.4.7) and (4.4.8), we get

$$|f(x_1) - f(x_2)| < \frac{4}{10n} < \frac{1}{2n},$$

contradicting (4.4.5). Thus, f must be of Baire class 1. That f is also a Darboux function follows from the conditions of the hypothesis [B2, p. 8].

To prove the converse, suppose f is a Darboux function of Baire class 1. Then there exists a sequence of continuous functions $\{f_k\}$ which converge pointwise to f on I_0. We shall assume that $I_0 = [0, 1]$ and that $f_1 \equiv 0$. We shall also assume that I denotes a closed subinterval of I_0. For each positive integer n, let

$$\mathbf{I}_n = \left\{ I : mI < \frac{1}{n} \text{ and } m(f_n(I)) < \frac{1}{n} \right\}.$$

Then each subinterval $I \in \mathbf{I}_1$. For each I, let

$$n(I) = \max\{n : I \in \mathbf{I}_n\}.$$

Now let I be any interval. Since f is a Darboux function, $f(I)$ is connected (Theorem 2.7.3). Since $f_{n(I)}$ is continuous, $f_{n(I)}(I)$ is a compact interval. Let $d(I)$ be the minimal distance between $\overline{f(I)}$ and $f_{n(I)}(I)$. Since $\overline{f(I)}$ is connected, we can choose $x_I \in I^\circ$ such that

$$|f_{n(I)}(x) - f(x_I)| < \frac{1}{n(I)} + d(I)$$

for all $x \in I$. Let $x_0 \in I$. From the preceding inequality, it follows that

$$|f(x_0) - f(x_I)| \leq |f(x_0) - f_{n(I)}(x_0)| + |f_{n(I)}(x_0) - f(x_I)|$$

$$< |f(x_0) - f_{n(I)}(x_0)| + \frac{1}{n(I)} + d(I)$$

$$\leq 2|f(x_0) - f_{n(I)}(x_0)| + \frac{1}{n(I)}.$$

Now fix x_0 and let $I \to x_0$. Then $n(I) \to \infty$. Thus, both terms following the last inequality sign approach zero as $I \to x_0$. Thus, $|f(x_0) - f(x_I)| \to 0$ as $I \to x_0$, which is what we wished to prove. $\qquad\square$

Lemma 4.4.9 [B2, p. 151]. *Let f be defined on an interval I_0. If f is the limit of a convergent interval function ϕ, then f is of Baire class 1.*

Proof. Suppose f is not in Baire class 1. Then, as in the previous theorem, we can find positive integers n and k, a perfect nonempty subset $Q \subset I_0$, and a set $Q_k \subset Q$ such that

(a) $\omega(f|_Q, x) \geq 1/n$ for all $x \in Q$,
(b) Q_k is dense in $Q \cap J$ for some interval J with $mJ < 1/k$,
(c) $|\phi(I) - f(x)| < 1/10n$ whenever $mI < 1/k$ and $x \in I^0 \cap Q_k$.

Now let I_1 be any interval contained in J such that $I_1 \cap Q$ is not empty and let $x_1 \in Q_k \cap I_1$. By (c),

(4.4.10) $$|\phi(I_1) - f(x_1)| < \frac{1}{10n}.$$

By (a), there exists $x_2 \in Q \cap I$ such that

(4.4.11) $$|f(x_1) - f(x_2)| > \frac{4}{10n}.$$

We can choose an interval $I_2 \subset I_1$ such that $x_2 \in I_2$ and

(4.4.12) $$|f(x_2) - \phi(I_2)| < \frac{1}{10n}$$

because f is the limit of the convergent interval function ϕ. Since Q_k is dense in Q, there exists $x_3 \in I_2 \cap Q_k$. Then

(4.4.13) $$|f(x_3) - \phi(I_2)| < \frac{1}{10n}$$

and

(4.4.14) $$|f(x_3) - \phi(I_1)| < \frac{1}{10n}.$$

because $m(I_2) < 1/k$ and $m(I_1) < 1/k$. From (4.4.12) and (4.4.13), we conclude

(4.4.15) $$|f(x_3) - f(x_2)| < \frac{2}{10n}.$$

From (4.4.11) and (4.4.15), we conclude

(4.4.16) $$|f(x_3) - f(x_1)| > \frac{2}{10n}.$$

From (4.4.10) and (4.4.14), we get

(4.4.17) $$|f(x_3) - f(x_1)| < \frac{2}{10n}.$$

Since (4.4.17) contradicts (4.4.16), our assumption that f was not of Baire class 1 was false. Thus, the lemma is proved.   □

Remark. In [B2, p. 151] the converse of Lemma 4.4.9 is also proved.

Theorem 4.4.18 [GN]. *Let $I_0 = [0, 1]$. Assume that $f: I_0 \to \mathbf{R}^1$ has an approximate derivative ADf everywhere on I_0. Then ADf is of Baire class 1.*

Proof. By Lemma 4.4.9, it suffices to show that ADf is the limit of a convergent interval function. For I, a subinterval of I_0, let

$$A(I, k) = \left\{ (x, y): x, y \in I \text{ and } \frac{f(y) - f(x)}{x - y} > k \right\}.$$

Define an interval function F by

$$F(I) = \sup \left\{ k: \frac{|A(I, k)|}{|I|^2} > \frac{1}{2} \right\},$$

where $|\cdot|$ denotes Lebesgue measure. We will show that F converges to ADf, i.e., for $x_0 \in I_0$ and $\{I_n\}$ a sequence of intervals in I_0 such that $x_0 \in \bigcap_{n \geq 1} I_n$ and $\lim |I_n| = 0$, we have

(4.4.19) $$\lim F(I_n) = ADf(x_0).$$

To prove (4.4.19), suppose $ADf(x_0) = \zeta$ and let $0 < \varepsilon < \frac{1}{8}$ be given. There is a positive integer N such that

(4.4.20) $$\left| \frac{f(x) - f(x_0)}{x - x_0} - \zeta \right| < \varepsilon$$

on a set E for which $|E \cap I_n|/|I_n| > 1 - \varepsilon$ whenever $n \geq N$.
    Let $n \geq N$ be fixed. For $x \in E_n = E \cap I_n$, let

$$E_x = \{ y: y \in E_n \text{ and } |x - x_0| < 8|x - y| \}.$$

We claim that

(4.4.21) $\qquad x \in E_n, \, y \in E_x \Rightarrow \left| \dfrac{f(y) - f(x)}{x - y} - \zeta \right| < 17\varepsilon.$

To prove (4.4.21), we may suppose $x > x_0$ and $y > x$ (the other cases being proved similarly). We may also suppose that $x_0 = 0$ and $f(x_0) = 0$. Then because $x \in E_n$ and $y \in E_x \subset E_n$, we have $x - x_0 < 8(y - x)$. Thus, $9x - x_0 < 8y$ which implies $9x - 9x_0 < 8y - 8x_0$. This, in turn, gives

$$\frac{x - x_0}{y - x_0} < \frac{8}{9}.$$

Let

$$0 < \frac{x - x_0}{y - x_0} = \tau < \frac{8}{9}.$$

Then by (4.18.2), we have

$$-\varepsilon < \zeta - \frac{f(x) - f(x_0)}{x - x_0} < \varepsilon$$

and

$$-\varepsilon < \frac{f(y) - f(x_0)}{y - x_0} - \zeta < \varepsilon.$$

Adding these two inequalities gives

$$-2\varepsilon < \frac{f(y) - f(x_0)}{y - x_0} - \frac{f(x) - f(x_0)}{x - x_0} < 2\varepsilon.$$

Therefore,

$$\frac{f(y) - f(x_0)}{y - x_0} - 2\varepsilon < \frac{f(x) - f(x_0)}{x - x_0} = \frac{f(x) - f(x_0)}{\tau(y - x_0)},$$

and, hence,

$$\tau \frac{f(y) - f(x_0)}{y - x_0} - 2\varepsilon\tau < \frac{f(x) - f(x_0)}{y - x_0},$$

or

$$\tau[f(y) - f(x_0)] - 2\varepsilon\tau(y - x_0) < f(x) - f(x_0).$$

Similarly,

$$f(x) - f(x_0) < \tau[f(y) - f(x_0)] + 2\tau\varepsilon(y - x_0).$$

Thus,

$$\frac{f(y) - f(x_0)}{y - x_0} - \frac{2\tau\varepsilon}{1 - \tau} < \frac{f(y) - f(x)}{y - x}$$

and

$$\frac{f(y) - f(x)}{y - x} < \frac{f(y) - f(x_0)}{y - x_0} + \frac{2\tau\varepsilon}{1 - \tau}.$$

Since $y \in E_x \subset E_n$, we have

$$\zeta - \varepsilon < \frac{f(y) - f(x_0)}{y - x_0} < \zeta + \varepsilon$$

and

$$\frac{2\tau}{1 - \tau} < 16.$$

Combining these results gives (4.4.21), i.e.,

$$\zeta - 17\varepsilon < \frac{f(y) - f(x)}{y - x} < \zeta + 17\varepsilon.$$

From the definition of E_n and E_x, we see that

(4.4.22) $x \in E_n \Rightarrow |E_x| > |I_n|(1 - \varepsilon)(\frac{3}{4}).$

To prove (4.4.22), suppose $x \in E_n$, $x > x_0$, $y \notin E_x$. Then $|x - x_0| \geq 8|x - y|$. Thus, either

$$x - x_0 \geq 8(x - y) \Rightarrow 8y \geq 7x + x_0 \Rightarrow y \geq \frac{7x}{8} + \frac{x_0}{8}$$

or

$$x - x_0 \geq 8(y - x) \Rightarrow 8y \leq 9x - x_0 \Rightarrow y \leq \frac{9x}{8} - \frac{x_0}{8};$$

that is, y is in the interval of length

$$\frac{2x}{8} - \frac{2x_0}{8} = \frac{x - x_0}{4},$$

i.e.,

$$|E_x^c| \leq |I_n|(1 - \varepsilon)(\frac{1}{4}).$$

From (4.4.21) and (4.4.22), the set of points $(x, y) \in I_n \times I_n$ for which

$$\left| \frac{f(y) - f(x)}{x - y} - \zeta \right| < 17\varepsilon$$

has measure $|I_n|^2(1 - \varepsilon)^2(\frac{3}{4}) > |I_n|^2/2$. From this it follows that $|F(I_n) - \zeta| < 17\varepsilon$ if $n \geq N$. Thus, $\lim F(I_n) = ADf(x_0)$. $\square$

Theorem 4.4.23 [PCP]. *Let $f: \mathbf{R}^1 \to \mathbf{R}^1$ be continuous. Then*

$$\{x: AD^-f(x) < D^+f(x) \text{ or } AD^+f(x) < D^-f(x)\}$$

is a set of the first category.

Proof. We will show that

$$A = \{x: AD^- f(x) < D^+ f(x)\}$$

is of the first category. For any rational number r, let

$$A_r = \{x: AD^- f(x) < r < D^+ f(x)\}.$$

Then $A = \bigcup_{r \in Q} A_r$. If $AD^- f(x) < r$, there is an integer N such that

$$m\left\{y: \frac{f(y) - f(x)}{y - x} > r, 0 < x - y < h\right\} \le \frac{h}{3}$$

for $h \in (0, 1/n)$. For $r \in Q$ and $n \in N$, define

$$\hat{A}_{rn} = \left\{x: m\left\{y: \frac{f(y) - f(x)}{y - x} > r, 0 < x - y < h\right\} \le \frac{h}{3} \text{ for } h \in \left(0, \frac{1}{n}\right)\right\}$$

and $A_{rn} = \hat{A}_{rn} \cap A_r$. Then $A_r = \bigcup \{A_{rn}: n \in N\}$ and $A = \bigcup \{A_{rn}: r \in Q, n \in N\}$. We will show that A_{rn} is nowhere dense. If possible, let $r \in Q$ and $n \in N$ be such that A_{rn} is not nowhere dense. Then there exists an open interval (α, β) with $0 < \beta - \alpha < 1/n$ such that A_{rn} is dense in (α, β); that is, $(\alpha, \beta) \subset \overline{A_{rn}} \subset \hat{A}_{rn}$.

If we show that $\hat{A}_{rn}$ is closed, then for fixed $x_0 \in (\alpha, \beta)$ and for any $x \in (x_0, \beta)$, we have $0 < x - x_0 < 1/n$ and $x \in A_{rn}$. This implies that $x \in \hat{A}_{rn}$. Choose $h = x - x_0$. Then

(4.4.24) $$m\left\{y: \frac{f(y) - f(x)}{y - x} > r, x_0 < y < x\right\} \le \frac{x - x_0}{3}.$$

Thus, there exists x_1, with $0 < x_1 - x_0 < \frac{1}{2}(x - x_0) < 1/2n \le \frac{1}{2}$ and

$$\frac{f(x_1) - f(x)}{x_1 - x} \le r.$$

Note that $x_0 + (x - x_0)/2 > x_0 + (x - x_0)/3$ and, thus, x_1 is a point between $x_0 + (x - x_0)/3$ and $x_0 + (x - x_0)/2$ not satisfying (4.4.24). But $x_1 \in (x_0, x) \subset (\alpha, \beta) \subset A_{rn}$ and $0 < x_1 - x_0 < \beta - \alpha < 1/n$. Now repeat the above argument and find x_2 such that $0 < x_2 - x_0 < (x_1 - x_0)/2 < (\frac{1}{2})^2$ and

$$\frac{f(x_2) - f(x_1)}{x_2 - x_1} \le r.$$

But this implies that

$$\frac{f(x_2) - f(x)}{x_2 - x} \le r$$

because x_2 will fall outside an interval of the correct length between x_0 and x_1. Continuing in this manner, we construct a sequence $\{x_n\} \in (\alpha, \beta)$ such that $x_n \to x_0$ with

$$\frac{f(x_n) - f(x)}{x_n - x} \le r.$$

Since f is continuous on $\mathbf{R}^1$, we get

$$\frac{f(x_0) - f(x)}{x_0 - x} \le r,$$

which implies that $D^+ f(x_0) \le r$. This means that $x_0 \notin A_r$ and hence $x_0 \notin A_{rn}$. But x_0 was any arbitrary point of (α, β). Thus, $(\alpha, \beta) \cap A_{rn}$ is empty, which is a contradiction to our assumption that A_{rn} is dense in (α, β).

Now let us prove that $\hat{A}_{rn}$ is closed by showing that its complement is open. To this end, we have to show that if $x_0 \notin \hat{A}_{rn}$, then there is a neighborhood of x_0 which is contained in $\hat{A}_{rn}^c$. If $x_0 \notin \hat{A}_{rn}$, there exists $h_0 \in (0, 1/n)$ such that

(4.4.25) $\qquad m\left\{y: \dfrac{f(y) - f(x_0)}{y - x_0} > r, 0 < x_0 - y < h_0\right\} > \dfrac{h_0}{3}.$

We are going to show that there exists $\varepsilon > 0$ such that for $x \in (x_0 - \varepsilon, x_0 + \varepsilon)$, we have

(4.4.26) $\qquad m\left\{y: \dfrac{f(y) - f(x)}{y - x} > r, 0 < x - y < h_0\right\} > \dfrac{h_0}{3},$

thereby implying that this entire ε-neighborhood is not in $\hat{A}_{rn}$. From (4.4.25) it follows that there exists $\delta > 0$ such that

$$\delta < \min\left(m\left\{y: \frac{f(y) - f(x_0)}{y - x_0} > r, 0 < x_0 - y < h_0\right\} - \frac{h_0}{3}, h_0\right).$$

But

$$\delta \ge m\left\{y: \frac{f(y) - f(x_0)}{y - x_0} > r, 0 < x_0 - y < h_0\right\}$$

$$- m\left\{y: \frac{f(y) - f(x_0)}{y - x_0} > r, \delta \le x_0 - y < h_0\right\}.$$

Combining these last two inequalities, we get

$$m\left\{y: \frac{f(y) - f(x_0)}{y - x_0} > r, \delta \le x_0 - y < h_0\right\}$$

$$\ge m\left\{y: \frac{f(y) - f(x_0)}{y - x_0} > r, 0 < x_0 - y < h_0\right\} - \delta > \frac{h_0}{3}.$$

Consider

$$F(x, y) = \frac{f(y) - f(x)}{y - x}$$

defined for $y < x$. Then F is continuous on $H = \{(x, y) \in \mathbf{R}^2 : y < x\}$ by continuity of f. Let $S = \{(x, y) \in H : |x - x_0| \le 1, \delta \le x - y \le h_0\}$. Then S is a compact set; hence, F is uniformly continuous on S. Now

$$\left\{y: \frac{f(y) - f(x_0)}{y - x_0} > r, \delta \le x_0 - y \le h_0\right\}$$

$$= \{y: F(x_0, y) > r, (x_0, y) \in S\}$$

$$= \bigcup_{k \in N} \left\{y: F(x_0, y) > r + \frac{1}{k}, (x_0, y) \in S\right\}.$$

Thus,

$$\lim_{k \to \infty} m\left\{y: (x_0, y) \in S, F(x_0, y) > r + \frac{1}{k}\right\}$$

$$= m\left\{y: \frac{f(y) - f(x_0)}{y - x_0} > r, \delta \le x_0 - y \le h_0\right\} > \frac{h_0}{3}.$$

This implies that there exists an integer k such that

$$m\left\{y: (x_0, y) \in S, F(x_0, y) > r + \frac{1}{k}\right\} > \frac{h_0}{3}.$$

By the uniform continuity of F on S, there exists $\eta \in (0, 2)$ such that if $(x_1, y_1) \in S$ and $(x_2, y_2) \in S$ with distance between (x_1, y_1) and (x_2, y_2) less than η, then

$$|F(x_1, y_1) - F(x_2, y_2)| < \frac{1}{k}.$$

Let $\varepsilon = \frac{1}{2}\eta$. Then if $(x_0, y) \in S$, $|x - x_0| < \varepsilon$, and $F(x_0, y) > r + 1/k$, we have $(x_0, y + (x - x_0)) \in S$ and the distance between (x_0, y) and $(x, y + (x - x_0))$ is less than or equal $2|x - x_0| < 2\varepsilon = \eta$. Since $F(x_0, y) > r + 1/k$, we use the uniform continuity criterion to get $F(x, y + (x - x_0)) > r$. Thus, for $x \in (x_0 - \varepsilon, x_0 + \varepsilon)$, we have

$$m\{y: (x, y) \in S, F(x, y) > r\} \ge m\left\{y: (x_0, y) \in S, F(x_0, y) > r + \frac{1}{k}\right\} > \frac{h_0}{3}.$$

This implies that for $x \in (x_0 - \varepsilon, x_0 + \varepsilon)$,

$$m\left\{y: \frac{f(y) - f(x)}{y - x} > r, 0 < x - y < h_0\right\}$$

$$= m\left\{y: \frac{f(y) - f(x)}{y - x} > r, 0 < x - y \le h_0\right\}$$

$$\ge m\left\{y: \frac{f(y) - f(x)}{y - x} > r, \delta \le x - y \le h_0\right\}$$

$$= m\{y: (x, y) \in S, F(x, y) > r\} > \frac{h_0}{3}.$$

Thus, $\hat{A}_{rn}^c$ is open.

We have shown that A_{rn} is nowhere dense and $\hat{A}_{rn}$ is closed. Thus, $A = \{x: AD^-f(x) < D^+f(x)\}$ is of the first category. Similarly, the set $\{x: AD^+f(x) < D^-f(x)\}$ is also of the first category.   □

Corollary 4.4.27 [PCP]. *The set*

$$\{x: D^-f(x) \neq D^+f(x) \text{ or } D_-f(x) \neq D_+f(x)\}$$

is of the first category.

Proof. The set $\{x: D^-f(x) \neq D^+f(x)\}$ is a subset of the set $\{x: AD^-f(x) < D^+f(x) \text{ or } AD^+f(x) < D^-f(x)\}$. So $\{x: D^-f(x) \neq D^+f(x)\}$ is a set of the first category. Since $D_+f(x) = -D^+(-f(x))$ and $D_-f(x) = -D^-(-f(x))$, we see that $\{x: D_-f(x) \neq D_+f(x)\}$ is also a set of the first category.   □

Theorem 4.4.28 [PCP]. *The set*

$$\{x: AD^+f(x) \neq AD^-f(x) \text{ or } AD_+f(x) \neq AD_-f(x)\}$$

is of the first category.

Proof. From Theorem 4.4.23 and the inequalities

$$D^+f(x) \geq AD^+f(x),$$
$$D^-f(x) \geq AD^-f(x)$$

(Exercise 5), we conclude that the set $\{x: AD^+f(x) \neq AD^-f(x)\}$ is of the first category. Since $AD_+f(x) = -AD^+(-f(x))$ and $AD_-f(x) = -AD^-(-f(x))$ (Exercise 6), it follows that $\{x: AD_+f(x) \neq AD_-f(x)\}$ is a set of the first category.

Corollary 4.4.29 [PCP]. *The set*

$$\{x: AD_+f(x) > AD^-f(x) \text{ or } AD_-f(x) > AD^+f(x)\}$$

is of the first category.

§4.5. Other Properties of Approximate Derivatives

In this section we consider results for approximate derivatives analogous to those in Chapter 3 for Dini derivatives in terms of monotonocity, the Darboux property, and so forth.

Theorem 4.5.1 [GN]. *Let $f: I_0 \to \mathbf{R}^1$ have an approximate derivative ADf everywhere on I_0. If $ADf \geq 0$ on I_0, then f is increasing on I_0.*

Proof. First suppose $ADf > 0$ on I_0. Let

$$E = \{x: x \in I_0 \text{ and } f(x) \geq f(0)\}.$$

Suppose $0 < \alpha < 1$ and let C be a set satisfying the following conditions:

(a) $C \subset E$,
(b) $x', x'' \in C$ and $x' < x''$ implies $|E \cap [x', x'']|/(x'' - x') \geq \alpha$.

Such sets C exist because $ADf > 0$. Let $\mathfrak{R}$ be the collection of all such sets C with partial ordering done by set inclusion. We see that every linearly ordered subset of $\mathfrak{R}$ has an upper bound in $\mathfrak{R}$, namely the union of the members of the subset. By Zorn's lemma, $\mathfrak{R}$ has a maximal element K.

Let $\beta = \sup K$. We will show that $\beta \in K$. If $x \in K$ and $x < \beta$, then there exists a sequence $\{x_n\}$ in K such that $x < x_n \leq x_{n+1}$ and $\lim x_n = \beta$. Since $K \in \mathfrak{R}$,

$$\alpha \leq \lim \frac{|E \cap [x, x_n]|}{x_n - x} = \frac{|E \cap [x, \beta]|}{\beta - x}.$$

But this implies β is a point of positive upper density of E. Note that f is approximately continuous at β. Hence, $x_n \in K \subset E$, implying $f(x) \geq f(0)$, i.e., $\beta \in E$.

Thus, we have shown that if $x \in K$, $x < \beta$, then

$$\frac{|E \cap [x, \beta]|}{\beta - x} \geq \alpha$$

and $\beta \in E$. Then $\beta \in K$ by the maximality of K.

Now we will show $\beta = 1$. If $\beta < 1$, there exists $\gamma > \beta$ such that $\gamma \in E$ and

$$\frac{|E \cap [\beta, \gamma]|}{\gamma - \beta} \geq \alpha.$$

This would imply that $\gamma \in K$ (because for every $x \in K$, $|E \cap [x, \gamma]| = |E \cap [x, \beta]| + |E \cap [\beta, \gamma]|$). This contradicts the maximality of K. Thus, $\beta = 1$ and $f(1) \geq f(0)$.

If $ADf \geq 0$, then for every $\varepsilon > 0$, $h_\varepsilon(x) = f(x) + \varepsilon x$ has a positive approximate derivative on I_0. Hence, $f(0) \leq f(1) + \varepsilon$. Since ε is arbitrary, we have $f(0) \leq f(1)$. Applying this argument to any $[x', x''] \subset I_0$, we get $f(x') \leq f(x'')$. $\qquad\square$

Theorem 4.5.2 [GN]. *Assume $f: I_0 \to \mathbf{R}^1$ has an approximate derivative ADf everywhere on I_0. Then ADf has the Darboux property, i.e., if $0 \leq \alpha < \beta \leq 1$ and η is between $ADf(\alpha)$ and $ADf(\beta)$, there exists $\xi \in (\alpha, \beta)$ such that $ADf(\xi) = \eta$.*

Proof. Suppose that ADf does not have the Darboux property. We can assume that $ADf(0) < 0$, $ADf(1) > 1$, and that there is no $\xi \in (0, 1)$ such that $ADf(\xi) = 0$. Let $E^+ = \{x : ADf(x) > 0\}$ and $E^- = \{x : ADf(x) < 0\}$. Then $I_0 = E^+ \cup E^-$.

We claim that if I is a component of either E^+ or E^-, then I is either a single point or a closed interval. To prove this, suppose I is not a single point and I is a component of E^+. Then I is an interval with endpoints *a* and *b*. By Theorem 4.5.1, f is increasing on (a, b). Thus, by approximate continuity of f, f is increasing on $[a, b]$. Then $ADf(a) \geq 0$ and $ADf(b) \geq 0$. Since $I_0 = E^+ \cup E^-$, we conclude that $ADf(a) > 0$ and $ADf(b) > 0$. Thus, $I = [a, b]$.

We may suppose that $0 \in E^-$ and $1 \in E^+$, by considering, if necessary, a subinterval of I_0. Let $\{I^+\}$ be the component intervals of E^+ and $\{I^-\}$ be those intervals of E^-. Let $\{I\} = \{I^+\} \cup \{I^-\}$. Then two distinct elements in $\{I\}$ are disjoint. Hence,

$$P = I_0 - \bigcup_{I \in \{I\}} I^0$$

is perfect and ADf has no point of continuity in P relative to P. Since this contradicts Theorem 4.4.18, our assumption was false.   □

Theorem 4.5.3 [GN]. *Let* $f: I_0 \to \mathbf{R}^1$ *have an approximate derivative everywhere on* I_0. *Let* $E = \{x: f'(x) \text{ exists}\}$. *Then for every subinterval* $I \subset I_0, I \cap E$ *contains an interval.*

Proof. Suppose there exists a subinterval $I \subset I_0$ such that $E \cap I$ contains no interval. We may assume that $I = I_0$. Let $E^+ = \{x: ADf(x) \geq 0\}$ and $E^- = \{x: ADf(x) < 0\}$. By Exercise 2, every component of E^+ and E^- is a single point. Since $I_0 = E^+ \cup E^-$, both E^+ and E^- are dense in I_0.

Let $\alpha > 0$ and let $E_\alpha^- = \{x: ADf(x) \leq -\alpha\}$. Then E_α^- is dense in I_0. To prove this, consider the function

$$g(x) = f(x) + \alpha x.$$

Then g satisfies the hypothesis of f and $A = \{x: g'(x) \text{ exists}\}$ contains no interval. Hence, $E_g^- = \{x: ADg(x) \leq 0\}$ is dense in I_0. But $ADg(x) = ADf(x) + \alpha$; therefore, $E_g^- = E_\alpha^-$.

Since E^+ and E_α^- are dense in I_0, ADf has no point of continuity in I_0, contradicting Theorem 4.4.18.   □

Theorem 4.5.4 [GN]. *Let* $f: I_0 \to \mathbf{R}^1$ *have an approximate derivative* ADf *on* I_0. *Then* ADf *has the mean value property; that is, if* $0 \leq \alpha < \beta \leq 1$, *there is* $\xi \in (\alpha, \beta)$ *such that*

$$f(\beta) - f(\alpha) = ADf(\xi)(\beta - \alpha).$$

Proof. In Theorem 4.5.2, we have shown that ADf possesses the Darboux property. We will now show that the Darboux property implies the mean value property. (In [GN] it is shown that the Darboux property and the mean value property are equivalent for approximate derivatives.) Suppose $0 \leq \alpha < \beta \leq 1$ and $f(\alpha) = f(\beta)$. We must show there exists $\xi \in (\alpha, \beta)$ such that $ADf(\xi) = 0$. If $ADf > 0$ on (α, β), then, by Theorem 4.5.1, $f(\alpha) < f(\beta)$. Similarly, $ADf < 0$ is impossible. Hence, there exists $\eta', \eta'' \in (\alpha, \beta)$ such that $ADf(\eta') \geq 0$ and $ADf(\eta'') \leq 0$. Since ADf has the Darboux property, there exists a $\xi \in (\alpha, \beta)$ such that $ADf(\xi) = 0$.

EXERCISES

1. See (4.4.20). Prove that if $ADf(x_0) = \zeta$ and $\varepsilon > 0$, there is a positive integer N such that

$$\left| \frac{f(x) - f(x_0)}{x - x_0} - \zeta \right| < \varepsilon$$

on a set E for which $|E \cap I_n|/|I_n| > 1 - \varepsilon$ whenever $n \geq N$.

2. Under the hypothesis of Theorem 4.5.1, ADf is the ordinary derivative of f.

3. If $f: I_0 = [0, 1] \to \mathbf{R}^1$ has an approximate derivative ADf everywhere on I_0 and if there exists a derivative ϕ' such that $\phi'(x) \geq ADf(x)$ [or $\phi'(x) \leq ADf(x)$] on I_0, then ADf is the ordinary derivative of f on I_0.

4. Let f be approximately differentiable on I_0. Then there exists a dense open subset of I_0 on which f is differentiable. [**Hint**: Use the fact that f is of Baire class 1 and Exercise 3.]

5. Prove $D^+f(x) \geq AD^+f(x)$ and $D^-f(x) \geq AD^-f(x)$.

6. Prove $AD_+f(x) = -AD^+(-f(x))$ and $AD_-f(x) = -AD^-(-f(x))$.

7. The following definition of the upper right approximate derivate is from [S1, p. 220]. Show that this definition is equivalent to Definition 4.1.2.

   Let f be a real-valued function defined on $\mathbf{R}^1$. The *approximate upper right-hand limit* of f at the point x_0, denoted $\limsup \mathrm{ap}_{x \to x_0+} f(x)$, is the lower bound of the numbers y ($+\infty$ included) for which the set $\{x: F(x) > y, x > x_0\}$ has x_0 as a point of dispersion. The *upper right approximate derivate* is

   $$\limsup_{x \to x_0+} \mathrm{ap} \left(\frac{f(x) - f(x_0)}{x - x_0} \right).$$

8. Show that if a function is approximately differentiable at a point, then it is approximately continuous at that point.

CHAPTER 5

Additional Results on Derivatives

In elementary calculus, we learn that if a function has a non-negative derivative on an interval, then the function is increasing on that interval. The first result in this chapter weakens the hypothesis on the function and obtains the same result. We then generalize this result with hypotheses involving derivates and approximate derivatives. In Section 5.4, we introduce the Denjoy property: If a function is differentiable everywhere on a closed interval, then the set of points where the derivative is bounded by any two extended-real numbers is either empty or of positive measure. We present Clarkson's proof of this result followed by a similar result for approximate derivatives. We conclude the chapter with a property related to the Denjoy property; for this last result, we require the concept of metrically dense sets.

§5.1. Derivatives

Theorem 5.1.1 (Goldowsky–Tonelli Theorem [S1, p. 206]). *If a continuous function f has a derivative (finite or infinite) at each point of $\mathbf{R}^1$ except possibly on a denumerable set, and if $f' \geq 0$ a.e., then f is an increasing function.*

Proof. Let E be the set of points such that if $x \in E$, then f is not monotone in any neighborhood of x. Clearly, E is a closed set and f is increasing on every interval contained in E^c. We will now show that E is empty.

We will suppose that $E \neq \varnothing$ and arrive at a contradiction. For every positive integer n, let P_n be the set of points x for which the inequality $0 < x' - x < 1/n$ implies $f(x') - f(x) \leq -(x' - x)$. Similarly, let Q_n be the set of points x for which the inequality $0 < x' - x < 1/n$ implies $f(x') - f(x) \geq -2(x' - x)$. We see that the sets P_n and Q_n are closed (some may be empty) and that they cover all of $\mathbf{R}^1$ except, at most, the finite or denumerable set of points at which f' does not exist.

By Baire's theorem, E contains a part which is either

(a) a single point, or
(b) contained in one of the P_n, or
(c) contained in one of the Q_n.

Since E has no isolated points, case (a) is impossible. Let us consider case (b). Suppose there exists a positive integer n_0 and an open interval I such that $\varnothing \neq E \cap I \subset P_{n_0}$. We can suppose that the diameter of I is less than $1/n_0$. Since $f'(x) \geq 0$ a.e., the set P_{n_0} is nondense. Let $[a, b]$ be any interval contiguous to $E \cap I$. The function f is increasing on $[a, b]$; this contradicts the fact that since $a, b \in P_{n_0}$ and $b - a < 1/n_0$, we have

$$f(b) - f(a) \leq -(b - a) < 0.$$

We now consider case (c). In this case, there exists an open interval I such that $E \cap I \neq \varnothing$ and $E \cap I \subset Q_n$ for some n. But then $D^+f(x) \geq -2$ everywhere in I and $f'(x) \geq 0$ a.e. in I. This implies f is increasing on I; this is impossible since I contains points of E in its interior.

Since each of the three cases leads to a contradiction, we have proved the theorem. □

§5.2. Derivates

Theorem 5.2.1 [K2, p. 181]. *Suppose f is continuous throughout a closed interval $[a, b]$. Let $E = \{x \in [a, b]: \underline{D}f(x) \leq 0\}$. If $f(E)$ contains no interval, then f is increasing on $[a, b]$.*

Proof. Suppose that the result is not true. Then there exist α and β such that $a \leq \alpha < \beta \leq b$ and $f(\alpha) > f(\beta)$. Since $f(E)$ contains no interval, there must be a number λ such that $f(\alpha) > \lambda > f(\beta)$ and

(5.2.2)   if $x \in [a, b]$ and $f(x) = \lambda$, then $\underline{D}f(x) > 0$.

Since f is continuous on $[\alpha, \beta]$, then the set

$$[\alpha, \beta] \cap \{x \in [a, b]: f(x) = \lambda\}$$

is not empty. Let ξ be the upper bound of this set. Then $\alpha < \xi < \beta$, $f(\xi) = \lambda$, and $f(x) < \lambda$ if $\xi < x < \beta$. But this contradicts (5.2.2) and thus proves the theorem. □

Theorem 5.2.3 [K2, p. 181]. *Suppose f is continuous throughout a closed interval $[a, b]$ and the following two conditions hold:*

(a) *the set of points $x \in [a, b]$ for which $\underline{D}f(x) = -\infty$ is finite or enumerable*
(b) *the set $Z = \{x \in [a, b]: \underline{D}f(x) < 0\}$ has measure zero,*

then $f(b) \geq f(a)$.

Proof. Let $\varepsilon > 0$. Then for every positive integer r, there exists an open set Z_r containing Z for which

$$(5.2.4) \qquad\qquad m(Z_r) < \varepsilon 2^{-r}.$$

Define the function $Z_r(x) = m((a, x) \cap Z_r)$, then $Z_r(x)$ is an increasing function of x and

$$(5.2.5) \qquad Z_r(b) - Z_r(a) \leq m(Z_r), \quad r = 1, 2, \ldots.$$

Since Z_r is open, if $x \in Z_r$, we have $Z_r(x + h) - Z_r(x) = h$ provided $|h|$ is small enough. Because $Z \subset Z_r$, it follows that if $x \in Z$, then $Z_r'(x) = 1$ for $r = 1, 2,$ .... Let

$$Z(x) = \sum_{r=1}^{\infty} Z_r(x).$$

Then by (5.2.4) and (5.2.5), we have that $Z(x)$ is increasing on $[a, b]$ and

$$(5.2.6) \qquad\qquad Z(b) - Z(a) < \varepsilon$$

and

$$(5.2.7) \quad \text{if } x \in Z, \text{ then } \underline{D}Z(x) \geq \underline{D} \sum_{r=1}^{n} Z_r(x) = n, \qquad n = 1, 2, \ldots;$$

i.e., $Z'(x) = \infty$. Let

$$\theta(x) = f(x) + Z(x) + \varepsilon x.$$

By (5.2.7), if $x \in Z$ and $\underline{D}f(x) > -\infty$, then $\underline{D}\theta(x) = \infty$. Since $Z(x)$ is increasing, if $x \in Z^c \cap [a, b]$, then $\underline{D}\theta(x) \geq \varepsilon$. Thus, if θ is continuous on $[a, b]$, it follows from condition (a) and Theorem 5.2.1 that $\theta(b) \geq \theta(a)$. Then by (5.2.6), we have

$$(5.2.8) \quad f(b) - f(a) \geq Z(a) - Z(b) - \varepsilon(b - a) \geq -\varepsilon - \varepsilon(b - a).$$

If we show that $Z(x)$ is continuous, it will follow that θ is continuous. Let $\delta > 0$. By (5.2.4), there exists n such that

$$(5.2.9) \qquad\qquad \sum_{r=n+1}^{\infty} mZ_r < \frac{\delta}{2}.$$

If $a \leq x < y \leq b$, we have

$$m((a, x) \cap Z_r) + m((x, y) \cap Z_r) = m((a, y) \cap Z_r), \quad r = 1, 2, \ldots.$$

Since $Z(x)$ is continuous,

$$(5.2.10) \qquad 0 \leq Z(y) - Z(x) = \sum_{r=1}^{\infty} m((x, y) \cap Z_r).$$

Note $m((x, y) \cap Z_r) \leq y - x$ and $m((x, y) \cap Z_r) \leq mZ_r$; by (5.2.9) and (5.2.10), we have

$$0 \leq Z(y) - Z(x) \leq n(y - x) + \frac{\delta}{2}.$$

Thus, $Z(x)$ is continuous. Since (5.2.8) holds for arbitrary ε, we have $f(b) \geq f(a)$.

□

§5.3. Approximate Derivatives

Lemma 5.3.1 [T5]. *If f is finite and approximately continuous at all points of a perfect set P and if it possesses an approximate derivative (finite or infinite) at every point of this set, there exists a portion of P on which f is continuous with respect to P.*

Proof. It is clear that

$$P = Q \cup R,$$

where $Q = \{x \in P: ADf(x) > -1\}$ and $R = \{x \in P: ADf(x) < 1\}$. Let Q_n be the set of points $x \in Q$ for which the density of the set

$$\left\{ \xi: \frac{f(\xi) - f(x)}{\xi - x} < 1 \right\}$$

is greater than $\frac{1}{2}$ on each segment $[x, x']$, verifying the condition $|x - x'| < 1/n$. Similarly, define the sets R_n. Then

$$P = \left(\bigcup_{n=1}^{\infty} Q_n \right) \cup \left(\bigcup_{n=1}^{\infty} R_n \right).$$

Consider any set Q_n (the reasoning for R_n is analogous). Let θ be a limit point of Q_n. Then f is continuous at the point θ with respect to Q_n. If not, there exists a sequence of points $\xi_1, \xi_2, \ldots, \xi_k, \ldots$ such that $\xi_k \in Q_n$ $(k = 1, 2, \ldots)$, $\xi_k \to \theta$ and $f(\xi_k)$ possesses a limit (finite or infinite) different from $f(\theta)$. We may suppose that

$$\xi_1 < \xi_2 < \cdots < \xi_k < \cdots < \theta$$

and

$$\lim_{k \to \infty} f(\xi_k) > f(\theta).$$

It is clear that for all values of k, we can suppose that

$$f(\xi_k) - f(\theta) > \delta > 0$$

and

$$\xi_{k+1} - \xi_k < \min\left\{ \frac{\delta}{2}, \frac{1}{n} \right\}.$$

It then follows that

$$\frac{f(x) - f(\xi_k)}{x - \xi_k} > -1$$

for all x belonging to a certain set E_k that is contained in $[\xi_k, \xi_{k+1}]$ and whose density on this segment is greater than $\frac{1}{2}$. Then for each $x \in E_k$, we have

$$f(x) - f(\xi_k) > -(x - \xi_k) \geq -(\xi_{k+1} - \xi_k) > -\frac{\delta}{2}.$$

Therefore, $f(x) > f(\theta) + \delta/2$ for all $x \in \bigcup_{k=1}^{\infty} E_k$. As

$$\frac{m(\bigcup_{k=m}^{\infty} E_k)}{\theta - \xi_m} = \frac{\sum_{k=m}^{\infty} mE_k}{\theta - \xi_m} > \frac{\frac{1}{2}\sum_{k=m}^{\infty}(\xi_{k+1} - \xi_k)}{\theta - \xi_m} = \frac{1}{2},$$

the function f cannot be approximately continuous at θ, which is a contradiction. Thus, f is continuous on each of the closed sets $\overline{Q}_n$ and $\overline{R}_n$. The proof of the lemma follows by observing that

$$P = \left(\bigcup_{n=1}^{\infty} \overline{Q}_n\right) \cup \left(\bigcup_{n=1}^{\infty} \overline{R}_n\right). \qquad \square$$

Remark. In the preceding lemma, we suppose that the function f has an approximate derivative at all points of the set P. However, no changes will be required to the proof if we suppose that the approximate derivative exists at all points of P with the exception of a denumerable subset of P.

Lemma 5.3.2 [T5]. *If f is a measurable function which is continuous with respect to a set P at the point x, if*

$$\frac{m\{\xi : x < \xi \leq y, [f(\xi) - f(x)]/(\xi - x) > k\}}{y - x} > \delta > 0$$

or

$$\frac{m\{\xi : x < \xi \leq y, |[f(\xi) - f(x)]/(\xi - x) - k| < \varepsilon\}}{y - x} > \delta > 0$$

or

$$\frac{m\{\xi : x < \xi \leq y, [f(\xi) - f(x)]/(\xi - x) < k\}}{y - x} > \delta > 0,$$

where y, k, ε, and δ are constants, then the above relations hold for all the points in P which are in a small enough neighborhood of x.

Proof. The proof will be the same for all three inequalities; therefore, we consider only the second. Designate the numerator of the left side of the inequality as $\mu(x)$. It suffices to show that μ is a lower semicontinuous function with respect to the set P at the point x, i.e.,

$$\lim_{x' \in P, x' \to x} \mu(x') \geq \mu(x).$$

$\mu(x)$ is equal to the measure of the set of points ξ such that $x < \xi \leq y$ for which the points with coordinates $(\xi, f(\xi))$ are in the interior of the angle A_x situated to the right of the point $(x, f(x))$ and formed by the two half-lines issuing from the point $(x, f(x))$ with angular coefficients[1] $k + \varepsilon$ and $k - \varepsilon$.

It suffices to give a proof of this lemma in the case where there exist some points $x' \in P$ as near as we wish to the point x for which $\mu(x') < \mu(x)$. In considering all mutual positions of the angles A_x and $A_{x'}$, we see that the least favorable case for the proof is that where the angle $A_{x'}$ is interior to the angle A_x, so we will restrict our proof to this case. Since f is continuous at x with respect to P, the sides of the angle $A_{x'}$ tend respectively toward those of the angle A_x when $x' \to x$ and x' remains in the set P. Moreover, it is clear that the difference $\mu(x) - \mu(x')$ is equal to the measure of the set of points ξ, $x < \xi \leq y$, for which the corresponding points $(\xi, f(\xi))$ are found to be on the sides of the angle $A_{x'}$, namely in the interior of the angle A_x but exterior to the angle $A_{x'}$. It follows that the difference $\mu(x) - \mu(x')$ tends to zero at the same time as $f(x) - f(x')$; this proves the lemma.   □

Definition 5.3.3 [T6]. Suppose f is defined on a set $\mathscr{C}$ and $\theta \in \mathscr{C}$. Let p, q, and y be constants. Define

$$\mu(\theta, y) = \frac{m\{\xi \colon \theta < \xi < y, [f(\xi) - f(\theta)]/(\xi - \theta) > p\}}{y - \theta}$$

and

$$v(\theta, y) = \frac{m\{\xi \colon \theta < \xi < y, [f(\xi) - f(\theta)]/(\xi - \theta) < q\}}{y - \theta}.$$

Remark. By the previous lemma, if f is continuous on $\mathscr{C}$ and $\mu(\theta, y) > \delta$ or $v(\theta, y) > \delta$, where δ is a constant such that $0 < \delta < 1$, then either $\mu(x, y) > \delta$ or $v(x, y) > \delta$ for all $x \in \mathscr{C}$ which belong to a certain segment Δ containing the point θ.

For the remainder of this section, we suppose that f is approximately continuous on the interval $[a, b]$ and that it possesses an approximate derivative (finite or infinite) at all points of this interval with the possible exception of a denumerable set E.

Lemma 5.3.4 [T6]. *If f is continuous at all points of a perfect set Π with*

---

[1] Angular coefficients are the slopes of the lines containing the rays which form the sides of an angle.

respect to this set, then the sets

$$Q = \{x : x \in \Pi, AD_+ f(x) > p\}$$

and

$$R = \{x : x \in \Pi, AD^+ f(x) < q\},$$

where p and q are constants such that $p > q$, cannot be simultaneously every-where dense in Π.

Proof. Suppose that the lemma is false, i.e., that Q and R are both everywhere dense in Π. We shall see that in this case there exists a sequence of points

$$y_1 > y_2 > \cdots > y_n > \cdots, \quad y_i \in [a, b] \ (i = 1, 2, 3, \ldots),$$

and a sequence of segments

$$\Delta_1 \supset \Delta_2 \supset \cdots \supset \Delta_n \supset \cdots, \quad \Delta_i = [\alpha_i, \beta_i] \subset [a, b] \ (i = 1, 2, 3, \ldots)$$

such that

(5.3.5) $\alpha_i < \beta_i < y_i \quad (i = 1, 2, 3, \ldots),$

(5.3.6) $\lim_{n \to \infty} (y_n - \alpha_n) = 0,$

(5.3.7) $\Pi \cap \Delta_i \neq \varnothing \quad (i = 1, 2, 3, \ldots),$

(5.3.8) $[E - (Q \cup R)] \cap \left(\bigcap_{n=1}^{\infty} \Delta_i \right) = \varnothing,$

(5.3.9) $\mu(x, y_i) > \frac{1}{2} \quad \text{if } i \text{ is odd and } x \in \Pi \cap \Delta_i,$

(5.3.10) $\nu(x, y_j) > \frac{1}{2} \quad \text{if } j \text{ is even and } x \in \Pi \cap \Delta_j.$

In fact, let $x_1, x_2, \ldots, x_n, \ldots$ all be points in the set $E - (Q \cup R)$ and $\varepsilon_1, \varepsilon_2, \ldots,$ $\varepsilon_n, \ldots$ positive numbers which converge to zero. As Q is dense in Π, we can find a point θ_1 such that $a < \theta_1 < b$ and $\theta_1 \in Q$. Therefore, there exists a point y_1 such that $\theta_1 < y_1 \leq b$, $\mu(\theta_1, y_1) > \frac{1}{2}$, and $y_1 - \theta_1 < \varepsilon_1/2$. By virtue of Lemma 5.3.2 and the remark following Definition 5.3.3, we can find a segment $\Delta_1 = [\alpha_1, \beta_1]$ with center θ_1 possessing the following properties:

$a \leq \alpha_1 < \theta_1 < \beta_1 < y_1,$
$x_1 \notin \Delta_1,$
$\Pi \cup \Delta_1$ is a perfect set,
$\mu(x, y_1) > \frac{1}{2}$ for all $x \in \Pi \cap \Delta_1.$

It is evident that

(5.3.11) $0 < y_1 - \alpha_1 < \varepsilon_1.$

Let us define y_2 and Δ_2. Choose a point θ_2 such that $\alpha_1 < \theta_2 < \beta_2$ and $\theta_2 \in R$. Then there exists a point y_2 such that $\theta_2 < y_2 < \beta_1$, $\nu(\theta_2, y_2) > \frac{1}{2}$, and $y_2 - \theta_2 < \varepsilon_2/2$. Again using Lemma 5.3.2, we can find a segment $\Delta_2 = [\alpha_2, \beta_2]$

with center θ_2 possessing the following properties:

$$\alpha_1 < \alpha_2 < \theta_2 < \beta_2 < y_2,$$
$$x_2 \notin \Delta_2,$$
$$\Pi \cap \Delta_2 \text{ is a perfect set,}$$
$$v(x, y_2) > \tfrac{1}{2} \text{ for all } x \in \Pi \cup \Delta_2.$$

Then, clearly,

(5.3.12) $0 < y_2 - \alpha_2 < \varepsilon_2.$

It is also evident that $y_1 > y_2$, $\Delta_2 \subset \Delta_1$, $x_1 \notin \Delta_1 \cap \Delta_2$, and $x_2 \notin \Delta_1 \cap \Delta_2$. Note that y_1, y_2, Δ_1, and Δ_2 satisfy conditions (5.3.5), (5.3.7), (5.3.9), and (5.3.10) for $i = 1$ and $j = 2$, as well as the inequalities (5.3.11) and (5.3.12).

Continuing in this manner, we can obtain y_n and Δ_n for all values of n which satisfy conditions (5.3.5), (5.3.7), (5.3.9), and (5.3.10) as well as the conditions

$$\Delta_j \subset \Delta_{j-1},$$
$$0 < y_j - \alpha_j < \varepsilon_j,$$
$$x_j \notin \bigcap_{i=1}^{n} \Delta_j$$

for $1 \le j \le n$. It is then clear that y_n and Δ_n satisfy all the conditions (5.3.5)–(5.3.10).

As $0 < \beta_n - \alpha_n < y_n - \alpha_n$, it follows from (5.3.6) that

$$\lim_{n \to \infty} (\beta_n - \alpha_n) = 0.$$

Therefore, by (5.3.7), all the Δ_n ($n = 1, 2, \ldots$) have a common point which belongs to $\cap \Delta_n$ and also to the set Π. Let θ be this point. By (5.3.8), we have $\theta \notin E - (Q \cup R)$. There are three possible cases: $\theta \in Q$, $\theta \in R$, or else at the point θ the approximate derivative $ADf(\theta)$ exists and we have $q \le ADf(\theta) \le p$.

The first case is impossible. By (5.3.5) and (5.3.6) we have that $\theta < y_i$ ($i = 1, 2, \ldots$) and $\lim_{n \to \infty} y_n = 0$; therefore, if $2n$ is large enough, we have $\mu(\theta, y_{2n}) < \tfrac{1}{2}$. But from (5.3.9), we get $v(\theta, y_{2n}) < \tfrac{1}{2}$. These two inequalities are clearly incompatible by Definition 5.3.3.

In a similar manner, it can be shown that the second case is impossible.

In the third case, it follows from the inequality $p > q$ that $ADf(\theta)$ cannot equal both of the numbers p and q. Suppose that $ADf(x) \neq p$. Then there exists a number r such that $ADf(\theta) < r < p$. For all values of n large enough, in particular for large odd values of n, we will have

$$\frac{m\{\xi : \theta < \xi < y_n, [f(\xi) - f(\theta)]/(\xi - \theta) < r\}}{y_n - \theta} > \frac{1}{2}.$$

On the other hand, for these odd values of n, we have $\mu(\theta, y_n) > \tfrac{1}{2}$. But these last two inequalities are incompatible. Thus, our premise that Q and R are both dense leads to a contradiction; therefore, the lemma is proved. □

Lemma 5.3.13 [T6]. *If at all the points of perfect set P, except perhaps at points of a denumerable subset E, the approximate derivative ADf exists and ADf ≠ −∞, we can find a portion Π ⊂ P such that if x' and x" are any two points in Π, then*

$$\frac{f(x') - f(x'')}{x' - x''} > -k,$$

where k is a certain positive constant.

Proof. By Lemma 5.3.1, there exists a portion Q of the set P such that f is continuous on Q with respect to Q. The set $Q \cap E$ may be written in the form

$$Q \cap E = \bigcup_{n=1}^{\infty} R_n,$$

where each set R_n contains at most one point. Let E_n be the set of points $x \in Q$ for which the density of the set

$$\mathscr{C}_x = \left\{ \xi : \frac{f(\xi) - f(x)}{\xi - x} > -n \right\}$$

exceeds $\frac{1}{2}$ on each segment $[x, x']$ provided that $|x - x'| < 1/n$. It is clear that

$$Q = \left(\bigcup_{n=1}^{\infty} E_n \right) \cup \left(\bigcup_{n=1}^{\infty} R_n \right).$$

Suppose that x' and x'' belong to the set E_n and that $|x' - x''| < 1/n$. Then on the segment $[x'', x']$ the sets $\mathscr{C}_{x'}$ and $\mathscr{C}_{x''}$ have some common points. Let ξ be one of these points and suppose $x' > x''$. Then

$$f(x') - f(x'') = f(x') - f(\xi) + f(\xi) - f(x'')$$

$$> -n(x' - \xi) - n(\xi - x'') = -n(x' - x''),$$

hence

$$\frac{f(x') - f(x'')}{x' - x''} > -n.$$

It follows from the continuity of f on Q that this relation is true for every pair of points x' and x'' in the closed set $\overline{E_n}$ for which $|x' - x''| < 1/n$.

Consider a function g which is equal to f on $\overline{E_n}$ and linear on the intervals contiguous to $\overline{E_n}$. Let $-m$ be the smallest in absolute value of the negative integers which are the lower bounds of the angular coefficients of the chords in the intervals contiguous to $\overline{E_n}$ whose length surpasses $1/n$. The derived numbers of g are at least equal to the $\min(-n, -m)$; therefore,

$$\frac{g(x') - g(x'')}{x' - x''} \geq \min(-n, -m),$$

for x', $x'' \in [a, b]$. Therefore, a fortiori, we have

$$\frac{f(x) - f(x'')}{x' - x''} \geq \min(-n, -m)$$

for all x', $x'' \in \overline{E}_n$. By letting $\min(-n, -m) - 1 = -k$, we will see that the lemma is proved. In fact, taking into account the equality

$$Q = \left(\bigcup_{n=1}^{\infty} \overline{E}_n \right) \cup \left(\bigcup_{n=1}^{\infty} R_n \right),$$

we can conclude (by Baire's theorem) that there is a portion $\Pi \subset Q$ on which Q (and therefore P) coincides with one of the $\overline{E}_n$. □

The next lemma requires an important result from Chapter 6 on functions of bounded variation. As most students of elementary analysis will be familiar with this result, we do not interrupt the flow of this presentation to introduce bounded variation here. If the result is not known to the reader, he may defer this lemma and the theorem following it until after Chapter 6.

Lemma 5.3.14 [T6]. *Let g be an approximately continuous function on $[a, b]$. If g is of bounded variation on $[a, b]$, then g is continuous on $[a, b]$.*

Proof. By the remark following Theorem 6.1.15, a function of bounded variation can have only points of discontinuity of the first category. However, by the definition of approximate continuity, g cannot have such discontinuities. □

Theorem 5.3.15 [T6]. *If an approximately continuous function f has an approximate derivative (finite or infinite) at each point of an interval $[a, b]$ except possibly on a denumerable set, and if $ADf(x) \geq 0$ a.e., then f is continuous and increasing and the approximate derivative is the ordinary derivative where it exists.*

Proof. Suppose that the theorem is not true. Let P be the set of points such that if $x \in P$, then f is not increasing in any neighborhood of x. If P is not empty, then P is perfect. By Lemma 5.3.1, there exists a portion $\Pi_1 \subset P$ such that f is continuous with respect to the set Π_1. For $\varepsilon > 0$, by Lemma 5.3.4 (letting $p = -\varepsilon$ and $q = -2\varepsilon$), there exists a portion $\Pi_2 \subset \Pi_1$ such that the set $\{x : AD_+ f(x) > -\varepsilon\}$ or else the set $\{x : AD^+ f(x) < -2\varepsilon\}$ has no point in Π_2.

The first case is not possible. This is immediate if $m(\Pi_2) > 0$ because $ADf(x) > 0$ a.e. in $[a, b]$; thus, $AD_+ f(x) > 0$ a.e. in $[a, b]$. If $m(\Pi_2) = 0$, we see that Π_2 is nondense because it is closed and of measure zero. Then at the left endpoints of the intervals contiguous to Π_2, we have $AD_+ f(x) \geq 0 > -\varepsilon$ (we consider only those intervals situated between the upper and lower bounds of Π_2).

Therefore, we see that it is the second case which must hold, i.e.,

$$\{x: AD^+ f(x) < -2\varepsilon\} \cap \Pi_2 = \varnothing.$$

If so, $ADf(x) = -\infty$ on Π_2.

By Lemma 5.3.13, there exists a portion $\Pi \subset \Pi_2$ such that

$$\frac{f(x') - f(x'')}{x' - x''} > -k$$

for any two points $x', x'' \in \Pi$ and for a certain positive constant k.

In the intervals contiguous to Π situated between the bounds α and β of the set Π, the function f is increasing. Therefore, the preceding inequality is true for each pair of points between α and β. But this means that f is of bounded variation on $[\alpha, \beta]$. Thus, by the previous lemma, it is continuous. Also its upper right Dini derivative is a.e. positive and is not $-\infty$ at any point. It follows that f is increasing on $[\alpha, \beta]$. Therefore, Π is empty. But this is impossible if P is not empty. Thus, we have obtained a contradiction to our supposition that f is not increasing. $\qquad\square$

We leave it to the reader to show that if f is an increasing function, then the ordinary derivative exists and is equal to the approximate derivative at each point where the latter exists (cf. Exercise 2 of Chapter 4).

Corollary 5.3.16 [T6]. *If f is approximately continuous on $[a, b]$ and has an approximate derivative (finite or infinite) at all points of $[a, b]$, with the possible exception of a denumerable set, and if $ADf(x) = 0$ a.e., then f is constant on $[a, b]$.*

Proof. From the previous theorem, it follows that

$$f(x) - f(a) \geq 0$$

for $a \leq x \leq b$. By considering the function $-f$, we get

$$-f(x) - [-f(a)] \geq 0.$$

Combining the two inequalities yields $f(x) = f(a)$. $\qquad\square$

§5.4. The Denjoy Property

Theorem 5.4.1 [C1]. *If f is real-valued and differentiable everywhere on the closed interval $[a, b]$, then for any two real numbers α and β ($\alpha < \beta$), the set*

$$E(\alpha, \beta) = \{x \in [a, b]: \alpha < f'(x) < \beta\}$$

is empty or of positive measure.

Proof. Suppose that $f'(x) \geq \lambda$ for almost all $x \in [a, b]$. Let

$$g(x) = f(x) - \lambda x.$$

Then g' exists everywhere in $[a, b]$. Since $g'(x) = f'(x) - \lambda \geq 0$ a.e., by Theorem 5.1.1, g is increasing and $f'(x) \geq \lambda$ for all $x \in [a, b]$. If $f'(x) \leq \mu$ a.e., we can apply Theorem 5.1.1 to the function $f - \mu$ to conclude $f'(x) \leq \mu$ everywhere. Thus, if f is differentiable everywhere, then the least upper bound and the greatest lower bound of its derivative are not changed if we ignore sets of measure zero in evaluating these bounds. This proves the theorem for the case $\alpha = -\infty$ or $\beta = +\infty$; this is the first part of the proof.

Now suppose α and β are arbitrary and finite. Let $E = E(\alpha, \beta)$. Suppose E is not empty and $mE = 0$; we shall show this leads to a contradiction. Let

$$E_\alpha = \{x : f'(x) \leq \alpha\},$$

$$E_\beta = \{x : f'(x) \geq \beta\}.$$

Then

$$[a, b] = E \cup E_\alpha \cup E_\beta,$$

the three sets being disjoint.

We claim $E \subset E_\alpha^c \cap E_\beta^c$. If not, then there exists x_0 which is in E but not in E_α^c. Then x_0 is in the interior of some interval U and $U \cap E_\alpha = \varnothing$. Since $x_0 \in E$, $\alpha < f'(x_0) < \beta$. Thus, U contains some points where $f' < \beta$. Also $mU = 0$ since $U \subset E$. This contradicts the first part of the proof. Thus, $E \subset E_\alpha^c$; similarly, $E \subset E_\beta^c$.

The derivative f' is of Baire class 1; hence, if $A \subset [a, b]$ and if we consider the function f' restricted to the domain A, the points of discontinuity form a set of first category relative to A. Let $A = \bar{E}$, i.e., the closure of the set $E(\alpha, \beta)$. A closed set is of second category relative to itself. Thus, we will have a contradiction if we show that f' is discontinuous everywhere on $\bar{E}$. To see this, proceed as follows: Let $x_0 \in E$. Since $E \subset E_\alpha^c \cap E_\beta^c$, we have for any interval I which contains x_0

$$\sup_{x \in I} f'(x) \geq \beta,$$

$$\inf_{x \in I} f'(x) \leq \alpha.$$

Because f' possesses the Darboux property (Definition 2.7.1), we have

$$\sup_{x \in I \cap E} f'(x) = \beta,$$

$$\inf_{x \in I \cap E} f'(x) = \alpha.$$

Thus, we have proved that, considered on the domain E (and so a fortiori on $\bar{E}$), the function f' is discontinuous at each point of E. On $\bar{E}$, the saltus of $f' \geq \beta - \alpha$. Hence, f' is everywhere discontinuous on $\bar{E}$, which leads to a contradiction. This completes the proof. $\square$

Remark. Theorem 5.4.1 was first proved by Denjoy in the case where the derivative is finite everywhere. Thus, the result of this theorem is sometimes called the Denjoy property. As a consequence of f' possessing the Darboux property, the set $E(\alpha, \beta)$ must either be empty or contain a continuum of points. But the result of Theorem 5.4.1 does not follow directly from the Darboux property, as seen in the following example of a function which possesses the Darboux property without satisfying Theorem 5.4.1.

EXAMPLE 5.4.2 [C1]. Let C be the Cantor ternary set and let $\{T_n\}$, $n = 1, 2, \ldots$, be a sequence of linear transformations such that the sets $T_n(C)$ are disjoint and such that any subinterval of $[0, 1]$ contains some $T_n(C)$. Let T_1 be the identity function. Define g on C in such a way that g assumes all values in the interval $[0, 1]$. On $T_n(C)$, let $g(x) = g(T_n^{-1}(x))$. On all remaining points of the interval $[0, 1]$, let $g(x) = 1$. Then g possesses the Darboux property, but the set

$$\{x \in [0, 1]: \tfrac{1}{2} < g(x) < 1\}$$

will be nonempty and of measure zero.

Theorem 5.4.3 [M1]. *If f is real-valued and approximately differentiable in the closed interval $[a, b]$, then for any two real numbers α and β $(\alpha < \beta)$, the set*

$$E_{AD}(\alpha, \beta) = \{x \in [a, b]: \alpha < ADf(x) < \beta\}$$

is empty or of positive measure.

Proof. The method of the proof is essentially the same as that of Theorem 5.4.1. We first consider the case where $ADf(x) \geq \lambda$ [or $ADf(x) \leq \mu$] a.e. and apply Theorem 5.3.15 to conclude that $ADf(x) \geq \lambda$ [or $ADf(x) \leq \mu$] everywhere on $[a, b]$. This proves the result if either $\alpha = -\infty$ or $\beta = +\infty$. By Theorem 4.4.18, ADf is of Baire class 1; by Theorem 4.5.2, ADf has the Darboux property. Thus, the remainder of the proof of Theorem 5.4.1 holds for ADf.   □

§5.5. Metrically Dense

To present the final result of this chapter (Theorem 5.5.9), we first need some results on metrically dense sets (not to be confused with the metric density of a set).

Definition 5.5.1 [H2, p. 191]. A set E is said to be *metrically dense* at a point p (p is not necessarily in E) if, for every interval I such that p is in the interior of I, the set $I \cap E$ has positive measure. A set of points is *metrically dense in itself* when every point of the set is a point at which the set is metrically dense.

EXAMPLE 5.5.2 [H2, p. 191]. The set of rational numbers in the interval $[0, 1]$ is everywhere dense but nowhere metrically dense.

Theorem 5.5.3 [H2, p. 191]. *The set of points at which a set E is metrically dense is closed.*

Proof. If in every neighborhood of a point p, there are points at which E is metrically dense, then it is clear that E is metrically dense at the point p.   □

Definition 5.5.4 [H2, p. 68]. We will construct a *system of nets* on the (finite or infinite) interval $[a, b]$ as follows: Divide $[a, b]$ into m_1 parts such that the length of each part does not exceed some positive number δ_1. Let D_1 denote the set of intervals obtained. Introduce new points of division of $[a, b]$ to create a set of intervals D_2 such that m_2 ($> m_1$) intervals are determined, the length of these intervals being less than a positive number δ_2 ($< \delta_1$). Continuing in this manner, we construct a sequence $D_1, D_2, \ldots, D_n, \ldots$ of sets of intervals, such that the length of the intervals in D_n do not exceed δ_n, where $\{\delta_n\}$ is a decreasing sequence which converges to zero. In addition, we choose the intervals in D_n to be closed on the left and open on the right, except for the extreme right interval which is closed. The set D_n is called a *net* and the nets corresponding to $n = 1, 2, 3, \ldots$ are called a *system of nets fitted on the interval* $[a, b]$. The separate intervals in D_n are called the *meshes* of the net D_n.

For each $x \in [a, b]$, x is in one and only one mesh of each of the nets D_n. These meshes, for $n = 1, 2, 3, \ldots$, are a convergent sequence of intervals for which x is the limit point.

Theorem 5.5.5 [H2, p. 191]. *Let E be a measurable set. Those points of E at which E is not metrically dense form a set of measure zero.*

Proof. Construct a system of nets for the interval in which E is contained. Let D_{n_1} be the first of the nets which has one or more meshes, each of which intersected with E is a set of measure zero. Let D'_{n_1} be the set of meshes in D_{n_1} for which this is true. Let D_{n_2} ($n_2 > n_1$) be the first net which has meshes, not contained in D'_{n_1}, each of which intersected with E is a set of measure zero. Let D'_{n_2} denote those meshes in D_{n_2} for which this is true. Continuing in this manner, we define a sequence $D'_{n_1}, D'_{n_2}, \ldots$ each of which consists of meshes in $D_{n_1}, D_{n_2}, \ldots$, respectively, such that D'_{n_r} does not belong to $D'_{n_{r-1}}$ for any value of r. Every point $p \in E$ at which E is not metrically dense belongs to one of the meshes of the sets $D'_{n_1}, D'_{n_2}, \ldots$, each of which contains a set of points of E of measure zero. Since the set of all points of E contained in all these meshes is a set of measure zero, the result is obtained.   □

Theorem 5.5.6 [H2, p. 191]. *Any measurable set of positive measure is the union of a set of measure zero and a set which is metrically dense in itself.*

Proof. By the previous theorem, if we remove all the points at which a set is not metrically dense from a set E, we are left with a set H such that $mH = mE$ and H is metrically dense in itself.   □

Theorem 5.5.7 [H2, p. 192]. *A closed set E of positive measure is the union of a set of measure zero and a perfect closed set G such that $mG = mE$ and G is metrically dense in itself.*

Proof. If E is closed, we may add to H (given in the previous theorem) all its limit points (which are in E), to obtain a closed set $\bar{H} \subset E$, where $m\bar{H} = mE$. $\bar{H}$ has the same properties as H, and $\bar{H}$ has no isolated points; hence, it is a perfect set.   □

Theorem 5.5.8 [H2, p. 192]. *Any measurable set E of positive measure contains a nondense perfect set G, metrically dense in itself, such that $mE - mG$ is arbitrarily small.*

Proof. Any measurable set E contains a nondense closed component, whose measure differs from the measure of E by an arbitrarily small amount. This nondense closed component contains a perfect set, also nondense, which is metrically dense in itself, and of measure equal to the measure of the component.

Theorem 5.5.9 [H3]. *The set $E(\alpha, \beta)$ (of Theorem 5.4.1) gives rise to a set of nonoverlapping and nonabutting open subintervals $\{I_i\}$ in the space $[a, b]$ and a closed set G, which is the complementary set of $\{I_i\}$ with respect to $[a, b]$, such that $E(\alpha, \beta)$ is void in each I_i and metrically dense everywhere in G.*

Proof. We will show that $E = E(\alpha, \beta)$ is metrically dense in itself. Let $p \in E$ and let U_p be any neighborhood containing p in its interior. Since $U_p \cap E$ is not empty, $m(U_p \cap E) > 0$ (by Theorem 5.4.1). Thus, each point of E is a limit point of the set. Let $A = \bar{E}^c$, i.e., the complement of the closure of E. For each $p \in A$, it is possible to construct a largest open subinterval I_p relative to the space $[a, b]$ such that

(a) I_p contains p in its interior,
(b) the two endpoints of I_p belong to $\bar{E}$,
(c) all the interior points of I_p are in A.

In the case $a \in A$ (or $b \in A$), the corresponding U_a (or U_b) takes a (or b) as one of its endpoints, the other endpoint being in $\bar{E}$. Therefore, $E \cap I_p$ is empty for each $p \in A$. Thus, corresponding to the set A, there exists a set of open subintervals $\{I_p\}$, each of whose intersection with E is empty. Also, if $p, p' \in A$, $p \neq p'$, then $I_p \cap I_{p'}$ is empty. For if not, then there would be an interior point of I_p (or $I_{p'}$) which is also in $\bar{E}$; this is contradictory to our method of constructing the intervals $\{I_p\}$. For the same reason, I_p and $I_{p'}$ cannot be abutting. Therefore, the set of open subintervals $\{I_p\}$ corresponding to the set A is a countable set of nonoverlapping and nonabutting open subintervals $\{I_i\}$ in the space $[a, b]$. Let G be the complement (with respect to $[a, b]$) of the collection $\{I_i\}$. Then G is closed; E is metrically dense everywhere in G, since for any point $q \in E, m(U_q \cap E) > 0$ is satisfied for any arbitrary neighborhood of q. □

Remark. The result of Theorem 5.5.9 is also true for the set $E_{AD}(\alpha, \beta)$. See [PP2] and [PP3] for the proof as well as some other results on properties of approximate derivatives.

CHAPTER 6

Bounded Variation

The class of functions of bounded variation, as well as the class of absolutely continuous functions (Chapter 7), play an important role in the analysis of functions of several variables. Partial differential equations, calculus of variations, and other applied fields require an extensive study of such functions. In this chapter, we introduce bounded variation for functions of a single variable. For completeness, the first section of the chapter contains results found in many textbooks. We then depart from the standard presentation on bounded variation with Kestelman's theorem on the Banach indicatrix. A section on the Stieltjes integral provides an application on bounded variation and gives the background needed to prove that the space $BV[a, b]$, with a suitable topology, is the dual of $C[a, b]$. Other topologies on this space will be introduced in Chapter 9. We next introduce the concept of local bounded variation and local integrability, followed by some additional remarks on Fubini's theorem. The chapter concludes with an extensive collection of interesting and challenging exercises.

§6.1. Bounded Variation on Finite Intervals

Definition 6.1.1. If $[a, b]$ is a compact interval, a finite set of points

$$P = \{x_0, x_1, x_2, \ldots, x_m\}$$

satisfying the inequalities

$$a = x_0 < x_1 < x_2 < \cdots < x_m = b$$

is called a *partition* of $[a, b]$. The collection of all possible partitions of $[a, b]$ is denoted by $\mathscr{P}[a, b]$.

Definition 6.1.2. Let $f: [a, b] \to \mathbf{R}^1$. If $P = \{x_0, x_1, x_2, \ldots, x_m\}$ is a partition of $[a, b]$, let

$$\Delta f_k = f(x_k) - f(x_{k-1})$$

for $k = 1, 2, \ldots, m$. If

$$T_f[a, b] = \sup_{p \in \mathscr{P}[a, b]} \sum_{k=1}^{m} |\Delta f_k| < \infty,$$

then f is said to be of *bounded variation on* $[a, b]$ ($f \in BV[a, b]$) and $T_f[a, b]$ is called the *total variation* of f on $[a, b]$. When the interval $[a, b]$ is clear, we shall simply write $f \in BV$.

Many of the results which follow are easy consequences of the definition of bounded variation and we frequently omit the proofs. The interested reader may consult any standard reference on real analysis ([A2], [J2], [N1], etc.) for the details.

Theorem 6.1.3. *If* $f: [a, b] \to \mathbf{R}^1$ *is monotonic, then* $f \in BV$ *and*

$$T_f[a, b] = |f(b) - f(a)|.$$

Definition 6.1.4. A finite function f defined on an interval $[a, b]$ is said to satisfy a *Lipschitz condition* if there exists a constant K such that, for any two points $x, y \in [a, b]$,

$$|f(x) - f(y)| \le K |x - y|.$$

Theorem 6.1.5. *If* f *satisfies a Lipschitz condition on* $[a, b]$*, then* $f \in BV$.

Corollary 6.1.6. *If* f *has a derivative at every point* $x \in [a, b]$ *and if* f' *is bounded on* $[a, b]$*, then* $f \in BV$.

Proof. Use the mean value theorem to show that f satisfies a Lipschitz condition on $[a, b]$. $\qquad\square$

Theorem 6.1.7. *If* $f \in BV[a, b]$*, then* f *is bounded on* $[a, b]$ *and*

$$|f(x)| \le |f(a)| + T_f[a, b].$$

EXAMPLE 6.1.8 [J2, p. 532]. That continuity and/or boundedness is not a sufficient condition for bounded variation is seen in the following example. Let

$$f(x) = \begin{cases} x \sin \dfrac{1}{x}, & x \in (0, 1] \\[2mm] 0, & x = 0. \end{cases}$$

Then f is continuous and bounded, but $f \notin BV[0,1]$. If

$$x_k = \frac{1}{(k + \frac{1}{2})\pi},$$

then

$$f(x_k) = \frac{(-1)^k}{(k + \frac{1}{2})\pi},$$

so

$$T_f[0,1] > \sum_{k=1}^{n} \frac{2}{k\pi},$$

which is unbounded as $n \to \infty$.

Theorem 6.1.9. *If $f, g \in BV[a,b]$, so is $f + g$ and*

$$T_{f+g}[a,b] \leq T_f[a,b] + T_g[a,b].$$

Theorem 6.1.10. *If $f \in BV[a,b]$ and $c \in \mathbf{R}^1$, then $cf \in BV[a,b]$ and*

$$T_{cf}[a,b] = |c|\, T_f[a,b].$$

Remark. Theorems 6.1.9 and 6.1.10 tell us that $BV[a,b]$ is a linear space.

Theorem 6.1.11. *If $f, g \in BV[a,b]$, then $fg \in BV[a,b]$. If, in addition,*

$$|g(x)| \geq c > 0$$

for all $x \in [a,b]$, then $f/g \in BV[a,b]$.

Theorem 6.1.12. *Let $f \in BV[a,b]$ and let $c \in (a,b)$. Then $f \in BV[a,c]$ and $f \in BV[c,b]$ and*

$$T_f[a,b] = T_f[a,c] + T_f[c,b].$$

That is, if we consider the total variation as a function of the interval, this function is an *additive* function.

Corollary 6.1.13 [N1, p. 218]. *If it is possible to divide the interval $[a,b]$ into a finite number of subintervals, on each of which a real-valued function f is monotonic, then $f \in BV[a,b]$.*

EXAMPLE 6.1.14 [R1, p. 419]. The following example illustrates that the converse of the previous corollary is not true; that is, a function of bounded variation need not be monotonic on any subinterval of its domain.

Let $r_1, r_2, r_3, \ldots$ be an ordering of the rational numbers in $(0,1)$ and let

$0 < a < 1$. Define $f: [0, 1] \to \mathbf{R}^1$ by

$$f(x) = \begin{cases} a^k, & x = r_k \\ 0, & \text{elsewhere.} \end{cases}$$

we will show that

$$T_f[0, 1] = \frac{2a}{1-a}.$$

Let $x_1 < x_3 < \cdots < x_{2n-1}$ be a listing of $r_1, r_2, \ldots, r_n$. Let $x_0 = 0$ and $x_{2n} = 1$. For $k = 1, 2, \ldots, n-1$, select x_{2k} such that

$$x_{2k-1} < x_{2k} < x_{2k+1}.$$

Then $P = \{x_0, x_1, \ldots, x_{2n}\}$ is a partition of $[0, 1]$ and

$$T_f[0, 1] \geq \sum_{i=1}^{2n} |f(x_i) - f(x_{i-1})| = 2 \sum_{k=1}^{n} a^k.$$

Since this holds for $n = 1, 2, 3, \ldots$, we have

$$T_f[0, 1] \geq \frac{2a}{1-a}.$$

On the other hand, let $P = \{x_0, x_1, \ldots, x_n\}$ be any partition of $[0, 1]$. For $k = 1, 2, \ldots, n-1$, there is one and only one i such that $r_k \in (x_{i-1}, x_{i+1})$. For this i,

$$|f(x_i) - f(x_{i-1})| + |f(x_{i+1}) - f(x_i)|$$

$$\leq \sup_{x \in [x_{i-1}, x_i]} f(x) - \inf_{x \in [x_{i-1}, x_i]} f(x) + \sup_{x \in [x_i, x_{i+1}]} f(x) - \inf_{x \in [x_i, x_{i+1}]} f(x)$$

$$\leq 2 \sum_{k=1}^{n-1} a^k < \frac{2a}{1-a}.$$

Thus, $T_f[0, 1] \leq 2a/(1 - a)$.

Theorem 6.1.15. $f \in BV[a, b]$ *if and only if it is the difference of two increasing functions.*

Proof. By Theorems 6.1.9 and 6.1.10, the difference of two increasing functions is a $BV[a, b]$ function. Now suppose $f \in BV[a, b]$. Define

$$p(x) = \tfrac{1}{2}(T_f[a, x] + f(x)),$$

$$n(x) = \tfrac{1}{2}(T_f[a, x] - f(x)).$$

Then $f(x) = p(x) - n(x)$. Thus, we must show that p and n are increasing functions. Suppose $\alpha, \beta \in [a, b]$, $\alpha < \beta$. Then

$$p(\beta) - p(\alpha) = \tfrac{1}{2}\{T_f[\alpha, \beta] - T_f[a, \alpha] + f(\beta) - f(\alpha)\}$$
$$\geq \tfrac{1}{2}\{T_f[\alpha, \beta] - |f(\beta) - f(\alpha)|\} \geq 0.$$

Similarly,

$$n(\beta) - n(\alpha) = \tfrac{1}{2}\{T_f[\alpha, \beta] - T_f[a, \alpha] - f(\beta) + f(\alpha)\}$$
$$\geq \tfrac{1}{2}\{T_f[\alpha, \beta] - |f(\beta) - f(\alpha)|\} \geq 0.$$

Remark. We note that the representation of a function of bounded variation as the difference of two increasing functions is not unique. Theorem 6.1.15 allows us to extend many of the results on monotone functions in Chapter 1 to functions $f \in BV[a, b]$:

(a) the limits $f(x+)$ $(x \in [a, b))$ and $f(x-)$ $(x \in (a, b])$ exist;
(b) f can have at most a countable number of discontinuities;
(c) the difference between a function of bounded variation and its saltus function is a continuous function of bounded variation;
(d) f' exists and is finite a.e.;
(e) f' is summable.

Theorem 6.1.16 [N1, p. 220]. *Let $H = \{f\}$ be an infinite family of functions defined on $[a, b]$. If all functions of the family are bounded by one and the same number*

$$|f(x)| \leq K,$$

then, for any countable subset $E \subset [a, b]$, it is possible to find a sequence $\{f_n\} \subset H$ which converges at every point of the set E.

Proof. Let $E = \{x_k\}$. Then

$$\{f(x_1): f \in H\}$$

is bounded. By the Bolzano–Weierstrass theorem, we can extract a convergent subsequence, call it $f_1^1(x_1), f_2^1(x_1), f_3^1(x_1), \ldots$, such that

$$\lim_{n \to \infty} f_n^1(x_1) = A_1.$$

Now consider the sequence $\{f_n^1(x_2)\}$. This is a bounded sequence and hence has a convergent subsequence, call it $f_1^2(x_2), f_2^2(x_2), f_3^2(x_2), \ldots$, such that

$$\lim_{n \to \infty} f_n^2(x_2) = A_2.$$

Continuing in this manner, we get

$$f_1^1(x_1), f_2^1(x_1), f_3^1(x_1), \ldots \to A_1,$$
$$f_1^2(x_2), f_2^2(x_2), f_3^2(x_2), \ldots \to A_2,$$
$$\vdots$$
$$f_1^k(x_k), f_2^k(x_k), f_3^k(x_k), \ldots \to A_k,$$

where each sequence is a subsequence of the previous one. We form a sequence of the diagonal elements $\{f_n^n(x_n)\}$, $n = 1, 2, 3....$ This sequence converges at every point of the set E since, for fixed k, $\{f_n^n(x_k)\}$ $(n \geq k)$, is a subsequence of $\{f_n^k(x_k)\}$ and hence converges to A_k.

Theorem 6.1.17 [N1, p. 221]. *Let $F = \{f\}$ be an infinite family of increasing functions defined on $[a, b]$. If all members of the family are bounded by one and the same number*

$$|f(x)| \leq K,$$

then there is a sequence of functions $\{f_n\} \subset F$ which converges to an increasing function at every point of $[a, b]$.

Proof. Let

$$E = \{Q \cap [a, b]\} \cup \{a\},$$

where Q is the set of rational numbers. Apply the previous theorem to E to find a sequence $\{f^n\} \subset F$ such that

$$\lim_{n \to \infty} f^n(x_k)$$

exists and is finite for each $x_k \in E$. Now define

$$\psi(x_k) = \lim_{n \to \infty} f^n(x_k), \quad x_k \in E.$$

For $x \in [a, b] - E$, define

$$\psi(x) = \sup\{\psi(x_k): x_k \in E, x_k < x\}.$$

Note that ψ is an increasing function on $[a, b]$; thus, the set of points where ψ is discontinuous is at most countable. Now we will show that

$$\lim_{n \to \infty} f^n(x_0) = \psi(x_0)$$

at every point x_0 where ψ is continuous. Let $\varepsilon > 0$ and let $x_i, x_k \in E$ such that

$$x_k < x_0 < x_i$$

and

$$\psi(x_i) - \psi(x_k) < \frac{\varepsilon}{2}.$$

Select a positive integer n_0 such that for $n > n_0$,

$$|f^n(x_k) - \psi(x_k)| < \frac{\varepsilon}{2}$$

and

$$|f^n(x_i) - \psi(x_i)| < \frac{\varepsilon}{2}.$$

Since ψ is increasing, we have

$$\psi(x_k) \le \psi(x_0) \le \psi(x_i) < \psi(x_k) + \frac{\varepsilon}{2}$$

and

$$\psi(x_k) - f^n(x_k) \le |f^n(x_k) - \psi(x_k)| < \frac{\varepsilon}{2}.$$

Thus,

$$\psi(x_0) < \psi(x_k) + \frac{\varepsilon}{2} < f^n(x_k) + \frac{\varepsilon}{2} + \frac{\varepsilon}{2}.$$

Therefore,

$$\psi(x_0) - \varepsilon < f^n(x_k).$$

Similarly,

$$f^n(x_i) < \psi(x_0) + \varepsilon.$$

Thus, for $n > n_0$,

$$\psi(x_0) - \varepsilon < f^n(x_k) \le f^n(x_0) \le f^n(x_i) < \psi(x_0) - \varepsilon$$

since each f^n is increasing. Therefore,

$$\lim_{n \to \infty} f^n(x_0) = \psi(x_0)$$

if ψ is continuous at x_0.

Now let E' be the countable set of points in $[a, b]$ where ψ is not continuous. Apply Theorem 6.1.16 to the set E' and the family of functions $\{f^n\}$. This yields a subsequence, call it $\{f_n\}$, which converges at all points of E'. Since the original sequence converged at all points of $[a, b] - E'$, $\{f_n\}$ converges at all points of $[a, b]$. Letting

$$\phi(x) = \lim_{n \to \infty} f_n(x)$$

we have an increasing function which converges for each $x \in [a, b]$. $\square$

Theorem 6.1.18 (Helly's First Theorem [N1, p. 222]). *Let $F = \{f\}$ be an infinite family of functions defined on $[a, b]$. If all functions of the family and the total variation of all functions of the family are bounded by a single number*

$$|f(x)| \le K \qquad and \qquad T_f[a, b] \le K,$$

then there exists a sequence $\{f_n\} \subset F$ which converges at every point of $[a, b]$ to some $\phi \in BV[a, b]$.

Proof. *For every $f \in F$, define two functions*

$$V(x) = T_f[a, x],$$

$$D(x) = V(x) - f(x).$$

Then V and D are increasing functions. Also

$$\begin{aligned}|V(x)| &\le |V(b)| \le K, \\ |D(x)| &\le |V(x)| + |f(x)| \le 2K.\end{aligned}$$

Apply Theorem 6.1.17 to the family of functions $\{V\}$ to get a convergent subsequence $\{V_k\}$ such that

$$\lim_{k\to\infty} V_k(x) = \alpha(x),$$

where α is increasing. For every function V_k, there exist functions D_k and f_k. Apply Theorem 6.1.17 to $\{D_k\}$ to get a convergent subsequence $\{D_{k_i}\}$ such that

$$\lim_{i\to\infty} D_{k_i}(x) = \beta(x),$$

where β is an increasing function. Then the sequence $V_{k_i} - D_{k_i} = f_{k_i} \in F$ and converges to

$$\phi = \lim_{i\to\infty} f_{k_i} = \lim_{i\to\infty} V_{k_i} - \lim_{i\to\infty} D_{k_i} = \alpha - \beta.$$

Since ϕ is the difference of two increasing functions, $\phi \in BV[a,b]$. $\qquad\square$

Theorem 6.1.19. *Let $f \in BV[a,b]$. Define*

$$V(a) = 0,$$

$$V(x) = T_f[a,x], \quad x \in (a,b].$$

Then every point of continuity of f is also a point of continuity of V. The converse is also true.

Proof. From the definition of bounded variation and the partition $P = \{a = x_0, x_1 = b\}$, we get

$$|f(b) - f(a)| \le T_f[a,b].$$

Thus, if $x < y$,

$$|f(y) - f(x)| \le T_f[x,y] = V(y) - V(x),$$

and if $y < x$, then

$$|f(y) - f(x)| \le T_f[y,x] = V(x) - V(y).$$

Thus, continuity of V implies continuity of f.

Now assume f is continuous at $c \in [a,b)$ and let $\varepsilon > 0$. Then there exists $\delta > 0$ such that

$$0 < |x - c| < \delta \Rightarrow |f(x) - f(c)| < \frac{\varepsilon}{2}.$$

For this ε, there exists a $P \in \mathscr{P}[c,b]$ such that

$$P = \{c = x_0, x_1, x_2, \ldots, x_n = b\}$$

and

$$T_f[c,b] - \frac{\varepsilon}{2} < \sum_{k=1}^{n} |\Delta f_k|.$$

If $x_1 - x_0 \geq \delta$, add a point $x_{1/2}$ to P such that $x_{1/2} - x_0 < \delta$. Then

$$T_f[c,b] - \frac{\varepsilon}{2} < |f(x_1) - f(x_0)| + \sum_{k=2}^{n} |\Delta f_k|$$

$$\leq |f(x_1) - f(x_{1/2})| + |f(x_{1/2}) - f(x_0)| + \sum_{k=2}^{n} |\Delta f_k|$$

$$< |f(x_1) - f(x_{1/2})| + \frac{\varepsilon}{2} + \sum_{k=2}^{n} |\Delta f_k|.$$

Since $\{x_{1/2}, x_1, x_2, \ldots, x_n\}$ is a partition of $[x_{1/2}, b]$, we have

$$T_f[c,b] - \frac{\varepsilon}{2} < \frac{\varepsilon}{2} + T_f[x_{1/2}, b].$$

Hence,

$$T_f[c,b] - T_f[x_{1/2}, b] < \varepsilon.$$

But

$$T_f[c,b] - T_f[x_{1/2}, b] = T_f[c, x_{1/2}] = T_f[a, x_{1/2}] - T_f[a,c] = V(x_{1/2}) - V(c).$$

Thus, if $x_{1/2} - c < \delta$, then $V(x_{1/2}) - V(c) < \varepsilon$. Therefore, $V(c+) = V(c)$; hence, V is right continuous at c. Similarly, it can be shown that if $c \in (a,b]$, then $V(c-) = V(c)$ if f is continuous at c. □

Corollary 6.1.20 [N1, p. 223]. *Let f be continuous on $[a,b]$. Then $f \in BV$ if and only if it can be expressed as the difference of two increasing continuous functions.*

Definition 6.1.21. Let $f: [a,b] \to \mathbf{R}^1$. Let $N(y)$ be the number of times that f assumes the value y for $x \in [a,b]$; $N(y) = +\infty$ if the value y is assumed infinitely often. The function $N(y)$ is called the *Banach indicatrix*.

It was proved by Banach [N1, p. 225] that if f is a continuous function of bounded variation on $[a,b]$, then the Banach indicatrix is summable and

$$\int_{-\infty}^{+\infty} N(y)\,dy = T_f[a,b].$$

We will present a more general result, due to Kestelman [K1], for discontinuous functions of bounded variation (Theorem 6.1.36).

We have already noted that if $f \in BV[a, b]$, then its points of discontinuity are enumerable and the limits $f(x+)$ $(x \in [a, b))$ and $f(x-)$ $(x \in (a, b])$ exist. Let $\{x_n\}$ be an infinite sequence of distinct points in $[a, b]$ containing all the discontinuities of f. Let $U(x)$ be the closed interval having $f(x)$ and $f(x+)$ for endpoints and $L(x)$ be the closed interval having $f(x)$ and $f(x-)$ for endpoints. Then

(6.1.22)
$$\sum_n |U(x_n)| + \sum_n |L(x_n)|,$$

where $|\cdot|$ denotes outer measure, converges to a value S, the sum of all the saltuses of f in $[a, b]$.

Let N_r be a partition of $[a, b]$ and defined by the points

$$a = \xi_0^r < \xi_1^r < \cdots < \xi_{n_r}^r = b.$$

For $1 \le m \le n_r$, let $[u_m^r]$, $[u_m^r)$, $(u_m^r]$, and (u_m^r) denote, respectively, the intervals $[\xi_{m-1}^r, \xi_m^r]$, $[\xi_{m-1}^r, \xi_m^r)$, $(\xi_{m-1}^r, \xi_m^r]$, and (ξ_{m-1}^r, ξ_m^r). Let $E_m^r = f([u_m^r])$, and let g_m^r and G_m^r be the lower and upper bounds, respectively, of the set E_m^r. Let

$$\Delta_m^r = (g_m^r, G_m^r)$$

and

$$\omega_r = \sum_{m=1}^{n_r} |\Delta_m^r|,$$

i.e., the oscillation of f in N_r. Because $f \in BV[a, b]$, we can show that a sequence of partitions $\{N_r\}$ exists with the following properties:

(6.1.23)
$$\lim_{r \to \infty} \omega_r = T_f[a, b].$$

(6.1.24)   The points defining N_r include $x_1, x_2, \ldots, x_r$.

(6.1.25)   The points defining N_{r+1} include all those defining N_r.

(6.1.26)
$$\lim_{r \to \infty} \max(\xi_m^r - \xi_{m-1}^r) = 0.$$

From (6.1.22) and (6.1.24), it follows that

(6.1.27)
$$\lim_{r \to \infty} \left\{ \sum_{m=0}^{n_r-1} |U(\xi_m^r)| + \sum_{m=1}^{n_r} |L(\xi_m^r)| \right\} = S.$$

Lemma 6.1.28 [K1]. *The sets E_m^r are measurable.*

Proof. Let D be the set of numbers $f(x_n+)$, $f(x_n-)$ $(n = 1, 2, \ldots)$. Let p be any limit point of E_m^r. Then there exists a point y and a sequence of points y_n, all in $[u_m^r]$, such that

$$\lim_{n \to \infty} y_n = y$$

and

$$\lim_{n \to \infty} f(y_n) = p.$$

If $f(y) \neq p$, then y is a point of discontinuity of f. Thus, $f(x+)$ and $f(x-)$ both exist and one of them must be equal to p (since the discontinuities of f are simple). Hence, if $p \in D^C$, then $f(y) = p$ and $p \in E_m^r$. It follows that $D^C \cap E_m^r$ is closed relative to D^C. Since D is enumerable, D is measurable. Therefore, D^C is measurable and, hence, $D^C \cap E_r^m$ is measurable. Since

$$|D \cap E_m^r| = 0,$$

we have

$$E_m^r = (D \cap E_{m_r}) \cup (D^C \cap E_m^r)$$

is measurable.

Lemma 6.1.29 [K1]. *Let*

$$O_m^r = \Delta_m^r \cap (E_{m_r})^C.$$

Then

$$|O_m^r| \le \sum_{x_n \in [u_m^r)} |U(x_n)| + \sum_{x_q \in (u_m^r]} |L(x_q)|.$$

Proof. Let k be any number such that

(6.1.30) $g_m^r < k < G_m^r,$

i.e., $k \in \Delta_m^r$, and

(6.1.31) $k \in (E_m^r)^C.$

It will be sufficient to prove that

(6.1.32)   either n exists such that $x_n \in [u_m^r)$ and $k \in U(x_n)$,
                      or q exists such that $x_q \in (u_m^r]$ and $k \in L(x_q)$.

Suppose the assertion is false for $k = k_0$. By (6.1.31), either $f(\xi_{m-1}^r) < k_0$ or $f(\xi_{m-1}^r) > k_0$. Suppose the former is true. Let α be the lower bound of points $x \in [u_m^r]$ for which $f(x) > k_0$. Since $k_0 \notin U(\xi_{m-1}^r)$, it follows that $f(\xi_{m-1}^r +) < k_0$. Hence,

(6.1.33) $\xi_{m-1}^r < \alpha \le \xi_m^r$

and

(6.1.34) $f(\alpha-) \le k_0.$

If $\alpha = \xi_m^r$, then $f(\xi_m^r) > k_0$, which, together with (6.1.34), is inconsistent with $k_0 \notin L(\xi_m^r)$. Hence,

$$\xi_{m-1}^r < \alpha < \xi_m^r$$

and

(6.1.35) $$\max\{f(\alpha), f(\alpha+)\} \geq k_0.$$

By (6.1.34) and (6.1.35),

$$k_0 \in U(\alpha) \cup L(\alpha).$$

Similarly, the assumption that $f(\xi^r_{m-1}) > k_0$ leads to a contradiction. $\square$

Theorem 6.1.36 [K1]. *Let* $f \in BV[a, b]$. *Then*

$$\int_{-\infty}^{+\infty} N(y)\,dy + S = T_f[a, b].$$

Proof. Let

(6.1.37) $$\phi^r_m(y) = \begin{cases} 1, & y \in f([u^r_m)) \\ 0, & \text{otherwise.} \end{cases}$$

By Lemma 6.1.28,

(6.1.38) $$|\Delta^r_m| - |O^r_m| = |E^r_m| = \int_{-\infty}^{+\infty} \phi^r_m(y)\,dy.$$

Let

$$\phi^r(y) = \sum_{m=1}^{n_r} \phi^r_m(y).$$

By (6.1.38),

(6.1.39) $$\omega_r - \sum_{m=1}^{n_r} |O^r_m| = \int_{-\infty}^{+\infty} \phi^r(y)\,dy.$$

By Lemma 6.1.29,

(6.1.40) $$\omega_r \leq \int_{-\infty}^{+\infty} \phi^r(y)\,dy + \sum_{x_n \in (a, b]} |L(x_n)| + \sum_{x_n \in [a, b)} |U(x_n)|.$$

Recall that every subinterval of N_{r+1} is part of a subinterval of N_r; hence,

$$0 \leq \phi^r(y) \leq \phi^{r+1}(y).$$

Define

$$\phi(y) = \lim_{r \to \infty} \phi^r(y).$$

By (6.1.23), (6.1.39), and Lebesgue's Monotone Convergence Theorem, it follows that $\phi(y)$ is summable and

$$\int_{-\infty}^{+\infty} \phi(y)\,dy = \lim_{r \to \infty} \int_{-\infty}^{+\infty} \phi^r(y)\,dy.$$

By (6.1.40), we have

(6.1.41)
$$T_f[a,b] \leq S + \int_{-\infty}^{+\infty} \phi(y)\,dy.$$

To prove that

$$T_f[a,b] \geq S + \int_{-\infty}^{+\infty} \phi(y)\,dy,$$

define g_m'' and G_m'' to be the lower and upper bounds, respectively, of $f((u_m^r))$, and v_m^r to be the total variation of f in $[u_m^r]$. Then

$$v_m^r \geq |U(\xi_{m-1}^r)| + |L(\xi_m^r)| + G_m'' - g_m''$$

$$\geq |U(\xi_{m-1}^r)| + |L(\xi_m^r)| + |E_m^r|,$$

since all the points of E_m^r, with the possible exception of $f(\xi_{m-1}^r)$ and $f(\xi_m^r)$, belong to $[g_m'', G_m'']$. Hence,

(6.1.42) $T_f[a,b] = \sum_{m=1}^{n_r} v_m^r \geq \sum_{m=0}^{n_r-1} |U(\xi_m^r)| + \sum_{m=1}^{n_r} |L(\xi_m^r)| + \sum_{m=1}^{n_r} |E_m^r|.$

By (6.1.27), (6.1.38), and (6.1.42), we have, letting $r \to \infty$,

(6.1.43)
$$T_f[a,b] \geq S + \int_{-\infty}^{+\infty} \phi(y)\,dy.$$

Combining (6.1.41) and (6.1.43) gives

$$T_f[a,b] = S + \int_{-\infty}^{+\infty} \phi(y)\,dy.$$

All that remains is to prove that $\phi(y) = N(y)$. We consider three cases:

(a) If $N(y) = 0$, then $y \notin f([u_m^r])$ for all r and all m. Thus, by (6.1.37),

$$\phi_m^r(y) = \phi^r(y) = \phi(y) = 0.$$

(b) If $N(y) = t < +\infty$, then by (6.1.26), each solution of the equation $y = f(x)$ will lie in a different subinterval $[u_m^r)$ for all large r, provided $y \neq f(b)$. Thus,

$$\phi_m^r(y) = 1$$

for exactly t values of m, and

$$\phi_m^r(y) = 0$$

for the remaining values of m, i.e.,

$$\phi_r(y) = t$$

for all large values of r. Hence,

$$\phi(y) = t,$$

provided $y \neq f(b)$.

(c) If $N(y) = +\infty$, there exists t distinct solutions of the equation $y = f(x)$, no matter how large t is. Hence, as in case (b), $\phi'(y) \geq t$ for all large r, which implies that $\phi(y) = +\infty$. □

§6.2. Stieltjes Integral

As an application of bounded variation, we shall now consider a generalization of the Riemann integral called the *Stieltjes integral*. This integral involves two functions: the integrand f and the integrator g. The integral (when it exists) is denoted by

$$\int_a^b f(x) \, dg(x)$$

and we call it the *Stieltjes integral of f with respect to g*. We shall not present a complete treatment of the Stieltjes integral here; the interested reader may consult [A2] or [N1]. Our main interest is the case when f is continuous and g is of bounded variation, or vice versa.

Definition 6.2.1. A partition P' is said to be *finer* than a partition P (or P' is a *refinement* of P) if $P' \subset P$. The *norm* of a partition P is the length of the longest subinterval of P and is denoted by $\|P\|$.

Definition 6.2.2. Let f and g be finite functions defined on an interval $[a,b]$. Let

$$P = \{a = x_0, x_1, \ldots, x_n = b\}$$

be a partition of $[a,b]$. Choose a point $t_k \in [x_{k-1}, x_k]$ and form the sum

$$S(P, f, g) = \sum_{k=1}^n f(t_k)[g(x_k) - g(x_{k-1})].$$

The number I is the Stieltjes integral of f with respect to g, if for every $\varepsilon > 0$, there exists a partition P_ε of $[a,b]$ such that, for every partition P finer than P_ε and for every choice of points $t_k \in [x_{k-1}, x_k]$, we have

$$|S(P, f, g) - I| < \varepsilon.$$

Symbolically, we say $f \in R(g)$ on $[a,b]$.

Theorem 6.2.3. *We list here some properties of the Stieltjes integral which follow directly from the definition, among them the property of linearity in both the integrator and the integrand.*

(6.2.4) *If f_1, $f_2 \in R(g)$ on $[a,b]$ and c_1, $c_2 \in \mathbf{R}^1$, then $c_1 f_1 + c_2 f_2 \in R(g)$ on $[a,b]$ and*

$$\int_a^b (c_1 f_1 + c_2 f_2)(x) \, dg(x) = c_1 \int_a^b f_1(x) \, dg(x) + c_2 \int_a^b f_2(x) \, dg(x).$$

(6.2.5) *If $f \in R(g_1)$ and $f \in R(g_2)$ on $[a, b]$ and c_1, $c_2 \in \mathbf{R}^1$, then $f \in R(c_1 g_1 + c_2 g_2)$ on $[a, b]$ and*

$$\int_a^b f(x) \, d(c_1 g_1 + c_2 g_2)(x) = c_1 \int_a^b f(x) \, dg_1(x) + c_2 \int_a^b f(x) \, dg_2(x).$$

(6.2.6) *Let $c \in (a, b)$. If any two of the following integrals exists, then the third exists also, and*

$$\int_a^c f(x) \, dg(x) + \int_c^b f(x) \, dg(x) = \int_a^b f(x) \, dg(x).$$

(6.2.7) *If $f \in R(g)$ on $[a, b]$, then $g \in R(f)$ on $[a, b]$ and*

$$\int_a^b f(x) \, dg(x) + \int_a^b g(x) \, df(x) = f(b)g(b) - f(a)g(a).$$

This equation is known as the formula for integration by parts.

Theorem 6.2.8 [N1, p. 230]. *If f is continuous on $[a, b]$ and $g \in BV[a, b]$, then the integral*

$$\int_a^b f(x) \, dg(x)$$

exists.

Proof. We may suppose that g is increasing. Let

$$P = \{a = x_0, x_1, \ldots, x_n = b\}$$

be a partition of $[a, b]$. Let

$$m_k = \inf\{f(x) : x \in [x_{k-1}, x_k]\},$$
$$M_k = \sup\{f(x) : x \in [x_{k-1}, x_k]\}.$$

Let

$$s = \sum_{k=1}^n m_k [g(x_k) - g(x_{k-1})],$$

$$S = \sum_{k=1}^n M_k [g(x_k) - g(x_{k-1})].$$

S and s are the *upper* and *lower* sums, respectively, associated with the partition P. Then

(6.2.9) $s \leq S(P, f, g) \leq S$

for all choices of the points $t_k \in [x_{k-1}, x_k]$. Let P_0 be a new partition obtained by adding the point x' of P. Then x' is in some interval $[x_{j-1}, x_j]$. Let

$$m' = \inf\{f(x) : x \in [x_{j-1}, x']\},$$
$$m'' = \inf\{f(x) : x \in [x', x_j]\}.$$

Then $m_j \leq m'$ and $m_j \leq m''$. Thus,

$$m_j[g(x_j) - g(x_{j-1})] \leq m''[g(x_j) - g(x')] + m'[g(x') - g(x_{j-1})].$$

Therefore, if s_0 is the lower sum associated with the new partition P_0, then $s \leq s_0$. Similarly, if S_0 is the upper sum associated with the partition P_0, we find $S \geq S_0$. Thus, we conclude that the lower sum does not decrease and the upper sum does not increase when new points are added to a partition. Hence, if S is the upper sum associated with any partition of $[a, b]$ and s is the lower sum associated with any (possibly different) partition, then

$$s \leq S.$$

Let I be the least upper bound of all the lower sums. Then by (6.2.9),

$$|S(P, f, g) - I| \leq S - s.$$

Since f is uniformly continuous on $[a, b]$, for $\varepsilon > 0$, we can find $\delta > 0$ such that

$$|x' - x''| < \delta \Rightarrow |f(x') - f(x'')| < \frac{\varepsilon}{2[g(b) - g(a)]}.$$

Then for any partition whose norm is less than δ, we have

$$M_k - m_k < \frac{\varepsilon}{2[g(b) - g(a)]}.$$

Thus,

$$S - s < \frac{\varepsilon}{2[g(b) - g(a)]} \sum_{k=1}^{n} [g(x_k) - g(x_{k-1})] = \frac{\varepsilon}{2} < \varepsilon.$$

Therefore,

$$|S(P, f, g) - I| < \varepsilon.$$

Since ε is arbitrary,

$$I = \int_a^b f(x) \, dg(x). \qquad \square$$

Theorem 6.2.10 [N1, p. 231]. *If f is continuous on $[a, b]$ and if g has a Riemann integrable derivative at every point of $[a, b]$, then*

$$\int_a^b f(x) \, dg(x) = \int_a^b f(x) g'(x) \, dx.$$

Proof. From the hypotheses of the theorem, it is clear that both integrals exist. Let

$$P = \{a = x_0, x_1, \ldots, x_n = b\}$$

be an arbitrary partition of $[a, b]$. Apply the mean value theorem to get

$$g(x_k) - g(x_{k-1}) = g'(y_k)(x_k - x_{k-1})$$

for some $y_k \in (x_{k-1}, x_k)$. Now use the points y_k for the points t_k in our definition of $S(P, f, g)$ to get

$$S(P, f, g) = \sum_{k=1}^{n} f(y_k)[g(x_k) - g(x_{k-1})] = \sum_{k=1}^{n} f(y_k)g'(y_k)(x_k - x_{k-1}).$$

This is the Riemann sum for $f(x)g'(x)$. Taking the limits on both sides now yields the desired result. □

Theorem 6.2.11 [N1, p. 232]. *If f is continuous on $[a, b]$ and $g \in BV[a, b]$, then*

$$\left| \int_a^b f(x) \, dg(x) \right| \leq M(f) T_g[a, b],$$

where

$$M(f) = \max\{|f(x)|: x \in [a, b]\}.$$

Proof. This result follows easily from the definition of the Stieltjes integral. □

Theorem 6.2.12 [N1, p. 232]. *Let $g \in BV[a, b]$ and $\{f_n\}$ be a sequence of continuous functions on $[a, b]$ which converge uniformly to the continuous function f. Then*

$$\lim_{n \to \infty} \int_a^b f_n(x) \, dg(x) = \int_a^b f(x) \, dg(x).$$

Proof. Let

$$M(f_n - f) = \max\{|f_n(x) - f(x)|: x \in [a, b]\}.$$

Then

$$\left| \int_a^b f_n(x) \, dg(x) - \int_a^b f(x) \, dg(x) \right| \leq M(f_n - f) T_g[a, b].$$

Since $f_n(x) \to f(x)$, $M(f_n - f) \to 0$, and the conclusion follows. □

Theorem 6.2.13 (Helly's Second Theorem [N1, p. 233]). *Let f be a continuous function defined on the interval $[a, b]$ and let $\{g_n\}$ be a sequence of functions which converges to a finite g at every point of $[a, b]$. If*

$$T_{g_n}[a, b] < K$$

for all n, then

$$\lim_{n \to \infty} \int_a^b f(x) \, dg_n(x) = \int_a^b f(x) \, dg(x).$$

Proof. First, we will show that $T_g[a, b] \leq K$. Let

$$P = \{a = x_0, x_1, \ldots, x_m = b\}$$

be a partition of $[a, b]$ and let $\varepsilon > 0$. Then for each x_k, $k \leq m$, there exists a positive integer N_k such that

$$n \geq N_k \Rightarrow |g_n(x_k) - g(x_k)| \leq \frac{\varepsilon}{2m}.$$

Let $N = \max\{N_k : x \leq m\}$. Then for $n \geq N$,

$$|g(x_k) - g(x_{k-1})|$$

$$\leq |g(x_k) - g_n(x_k)| + |g_n(x_k) - g_n(x_{k-1})| + |g_n(x_{k-1}) - g(x_{k-1})|$$

$$\leq |g_n(x_k) - g_n(x_{k-1})| + \frac{\varepsilon}{m}.$$

Thus,

$$\sum_{k=1}^{m} |g(x_k) - g(x_{k-1})| \leq \sum_{k=1}^{m} |g_n(x_k) - g_n(x_{k-1})| + \varepsilon \leq T_{g_n}[a, b] + \varepsilon < K + \varepsilon.$$

Since ε and the partition are arbitrary, it follows that $T_g[a, b] \leq K$.

Now, for $\varepsilon > 0$, select a partition $P \in \mathscr{P}[a, b]$ such that the oscillation of f on each subinterval is less than $\varepsilon/3K$. Then

$$\int_a^b f(x)\, dg(x) = \sum_{k=1}^{m} \int_{x_{k-1}}^{x_k} f(x)\, dg(x)$$

$$= \sum_{k=1}^{m} \int_{x_{k-1}}^{x_k} [f(x) - f(x_k)]\, dg(x) + \sum_{k=1}^{m} \int_{x_{k-1}}^{x_k} f(x_k)\, dg(x)$$

$$< \sum_{k=1}^{m} \frac{\varepsilon}{3K} \int_{x_{k-1}}^{x_k} dg(x) + \sum_{k=1}^{m} f(x_k) \int_{x_{k-1}}^{x_k} dg(x)$$

$$= \sum_{k=1}^{m} \frac{\varepsilon}{3K} [g(x_k) - g(x_{k-1})] + \sum_{k=1}^{m} f(x_k)[g(x_k) - g(x_{k-1})]$$

$$\leq \frac{\varepsilon}{3K} K + \sum_{k=1}^{m} f(x_k)[g(x_k) - g(x_{k-1})].$$

Therefore,

$$\int_a^b f(x)\, dg(x) = \sum_{k=1}^{m} f(x_k)[g(x_k) - g(x_{k-1})] + \theta\left(\frac{\varepsilon}{3}\right), \quad |\theta| \leq 1.$$

Similarly,

$$\int_a^b f(x)\, dg_n(x) = \sum_{k=1}^{m} f(x_k)[g_n(x_k) - g_n(x_{k-1})] + \theta_n\left(\frac{\varepsilon}{3}\right), \quad |\theta_n| \leq 1.$$

Thus,

$$\left| \int_a^b f(x) \, dg_n(x) - \int_a^b f(x) \, dg(x) \right|$$

$$= \left| \sum_{k=1}^m f(x_k)[g_n(x_k) - g_n(x_{k-1})] \right.$$

$$\left. - \sum_{k=1}^m f(x_k)[g(x_k) - g(x_{k-1})] + \theta_n\left(\frac{\varepsilon}{3}\right) - \theta\left(\frac{\varepsilon}{3}\right) \right|$$

$$\leq M \sum_{k=1}^m |g_n(x_k) - g(x_k)| + |g(x_{k-1}) - g_n(x_{k-1})| + \frac{2\varepsilon}{3},$$

where

$$M = \max\{|f(x)|: x \in [a,b]\}.$$

For each x_k, $k \leq m$, there exists a positive integer N_k^* such that

$$n \geq N_k^* \Rightarrow |g(x_k) - g_n(x_k)| \leq \frac{\varepsilon}{6Mm}.$$

Let $N^* = \max\{N_k^*: k \leq m\}$. Then, if $n \geq N^*$, we have

$$M \sum_{k=1}^m [|g_n(x_k) - g(x_k)| + |g(x_{k-1}) - g_n(x_{k-1})|] \leq \frac{\varepsilon}{3}.$$

Hence,

$$\left| \int_a^b f(x) \, dg_n(x) - \int_a^b f(x) \, dg(x) \right| < \varepsilon$$

if $n \geq N^*$. Thus, we have proved the result. $\square$

§6.3. The Space $BV[a,b]$

We have already observed that the space $BV[a,b]$ is a linear space. We will now introduce a norm under which this space is complete. We will then show that every bounded linear functional on the space $C[a,b]$ can be represented as a Stieltjes integral.

Theorem 6.3.1. $BV[a,b]$ is a Banach space (a complete normed linear space) under the norm

$$\|f\| = |f(a)| + T_f[a,b].$$

Proof. The reader should first verify that $\|\cdot\|$ satisfies all the properties of a norm. Suppose $\{f_n\}$ is a Cauchy sequence in $BV[a,b]$; that is, if $\varepsilon > 0$, there exists a positive integer N, such that if $n, m \geq N$, then

$$\| f_n - f_m \| < \varepsilon,$$

i.e.,

$$|f_n(a) - f_m(a)| + T_{f_n - f_m}[a,b] < \varepsilon.$$

We note that this implies that $\{f_n(a)\}$ is a Cauchy sequence of real numbers. Let $x \in (a,b)$. Then $\{a, x, b\}$ is a partition of $[a,b]$. Therefore,

$$|f_n(x) - f_m(x) - \{f_n(a) - f_m(a)\}| + |f_n(b) - f_m(b) - \{f_n(x) - f_m(x)\}|$$
$$\leq T_{f_n - f_m}[a,b].$$

Thus,

$$|f_n(x) - f_m(x)| \leq |f_n(a) - f_m(a)| + T_{f_n - f_m}[a,b].$$

The terms on the right can be made arbitrarily small by taking n and m sufficiently large. Thus, $\{f_n(x)\}$ is a Cauchy sequence for each $x \in (a,b)$. By taking the partition $\{a,b\}$, we can show that $\{f_n(b)\}$ is also a Cauchy sequence of real numbers. Hence,

$$f(x) = \lim_{n \to \infty} f_n(x)$$

exists for each $x \in [a,b]$. Since $\{f_n\}$ satisfies the Cauchy condition for uniform convergence of a sequence of functions, for $\varepsilon > 0$ and any arbitrary partition $P = \{a = x_0, x_1, \ldots, x_K = b\}$ of $[a,b]$, there exists a positive integer N^* such that

$$|f_{N^*}(x) - f(x)| < \frac{\varepsilon}{2K}$$

for each $x \in [a,b]$. Then

$$\sum_{i=1}^{K} |f(x_i) - f(x_{i-1})|$$

$$\leq \sum_{i=1}^{K} \{|f(x_i) - f_{N^*}(x_i)| + |f_{N^*}(x_i) - f_{N^*}(x_{i-1})| + |f_{N^*}(x_{i-1}) - f(x_{i-1})|\}$$

$$< \sum_{i=1}^{K} \left\{ f_{N^*}(x_i) - f_{N^*}(x_{i-1})| + \frac{\varepsilon}{K} \right\} \leq V_{f_{N^*}}[a,b] + \varepsilon.$$

Since P and ε are arbitrary, it follows that

$$V_f[a,b] \leq V_{f_{N^*}}[a,b].$$

Thus, $f \in BV[a,b]$, which was what we wanted to prove. $\square$

Definition 6.3.2. A *linear functional* on a normed linear space V is a mapping

$$\phi: V \to \mathbf{R}^1$$

such that $\phi(\alpha f + \beta g) = \alpha\phi(f) + \beta\phi(g)$ for all f, $g \in V$ and all α, $\beta \in \mathbf{R}^1$. A linear functional is said to be bounded if there exists a real number $M > 0$ such that

$$|\phi(f)| \le M \|f\|_V$$

for all $f \in V$. The smallest M for which this inequality holds is called the norm of ϕ. Thus,

$$\|\phi\| = \sup \frac{|\phi(f)|}{\|f\|_V}$$

as f ranges over all nonzero elements of V. The set of all bounded linear functionals on a space V is called the *dual space* (*conjugate space, adjoint space*) of V and is denoted V^*.

Let $C[a,b]$ denote the space of functions which are continuous on $[a,b]$ with norm

$$\|f\| = \max\{|f(x)|: x \in [a,b]\}.$$

Then if $g \in BV[a,b]$, we can define a function

$$\phi: C[a,b] \to \mathbf{R}^1$$

by

$$\phi(f) = \int_a^b f(x)\, dg(x).$$

Using the properties of the Stieltjes integral, we can show

(1) $\phi(\alpha f_1 + \beta f_2) = \alpha\phi(f_1) + \beta\phi(f_2)$,
(2) $|\phi(f)| \le T_g[a,b] \|f\|$.

Thus, Stieltjes integrals of the above form are bounded linear functionals on the space $C[a,b]$. The following theorem, due to F. Riesz, shows that these are the only such functionals on this space.

Theorem 6.3.3. *Let ϕ be a bounded linear functional on the space $C[a,b]$. Then there exists $g \in BV[a,b]$ such that*

$$\phi(f) = \int_a^b f(x)\, dg(x)$$

for all $f \in C[a,b]$ and $\|\phi\| = T_g[a,b]$.

Proof. Let ϕ be a bounded linear functional on $C[a,b]$. By the Hahn–Banach theorem, we can extend ϕ to $L^\infty[a,b]$ without increasing its norm. (Recall $L^\infty[a,b]$ is the space of all measurable functions which are bounded except possibly on a set of measure zero. Thus, $C[a,b] \subset L^\infty[a,b]$.) For each

$x \in [a,b]$, define

$$u_x(t) = \begin{cases} 1, & a \leq t \leq x \\ 0, & t > x. \end{cases}$$

Let $g(x) = \phi(u_x)$. We will show that $g \in BV[a,b]$ and $T_g[a,b] \leq \|\phi\|$. Let $\{a = x_0, x_1, \ldots, x_n = b\}$ be an arbitrary partition of $[a,b]$. Let

$$\varepsilon_i = \text{sgn}[g(x_i) - g(x_{i-1})], \quad i = 1, \ldots, n.$$

Then

$$\sum_{i=1}^{n} |g(x_i) - g(x_{i-1})| = \sum_{i=1}^{n} \varepsilon_i[g(x_i) - g(x_{i-1})]$$

$$= \sum_{i=1}^{n} \varepsilon_i[\phi(u_{x_i}) - \phi(u_{x_{i-1}})]$$

$$= \phi\left[\sum_{i=1}^{n} \varepsilon_i(u_{x_i} - u_{x_{i-1}})\right]$$

$$\leq \|\phi\| \cdot \left\|\sum_{i=1}^{n} \varepsilon_i(u_{x_i} - u_{x_{i-1}})\right\| \leq \|\phi\|$$

because

$$\left\|\sum_{i=1}^{n} \varepsilon_i(u_{x_i} - u_{x_{i-1}})\right\| \leq 1.$$

Since the partition was arbitrary, it follows that $T_g[a,b] \leq \|\phi\|$ and $g \in BV[a,b]$. Thus,

$$\int_a^b f(x)\,dg(x)$$

is defined for all $f \in C[a,b]$. Now we will show that

$$\phi(f) = \int_a^b f(x)\,dg(x).$$

Suppose $f \in C[a,b]$. Let

$$x_i = a + \frac{i(b-a)}{n}, \quad i = 0, 1, \ldots, n.$$

Define

$$f_n(x) = \sum_{i=1}^{n} f(x_i)[u_{x_i}(x) - u_{x_{i-1}}(x)].$$

Then

$$\|f - f_n\| = \sup\{|f(x) - f_n(x)| : x \in [a,b]\}$$

$$= \max_{1 \leq i \leq n} \sup\{|f(x) - f(x_i)| : x_{i-1} < x \leq x_i\}.$$

Since f is uniformly continuous on $[a,b]$, we have $|f(x) - f(x_i)| \to 0$ as $n \to \infty$. Thus, $\|f - f_n\| \to 0$, and hence $f_n \to f$. Since a bounded linear functional is necessarily continuous, it follows that

$$
\begin{aligned}
\phi(f) = \lim_{n \to \infty} \phi(f_n) &= \lim_{n \to \infty} \phi\left[\sum_{i=1}^{n} f(x_i)\{u_{x_i}(x) - u_{x_{i-1}}(x)\} \right] \\
&= \lim_{n \to \infty} \sum_{i=1}^{n} f(x_i)[\phi(u_{x_i}) - \phi(u_{x_{i-1}})] \\
&= \lim_{n \to \infty} \sum_{i=1}^{n} f(x_i)[g(x_i) - g(x_{i-1})] \\
&= \int_a^b f(x)\,dg(x).
\end{aligned}
$$

Thus, we have

$$
|\phi(f)| \le T_g[a,b] \cdot \|f\|.
$$

Hence,

$$
\|\phi\| \le T_g[a,b].
$$

This completes the proof. $\qquad\qquad\qquad\qquad\qquad\qquad\qquad\qquad\qquad\square$

Remark. The preceding theorem makes no assertion about the uniqueness of the function $g \in BV[a,b]$. We can see that if

$$
\phi(f) = \int_a^b f(x)\,dg(x)
$$

then

$$
\phi(f) = \int_a^b f(x)\,dg_1(x),
$$

where $g_1(x) = g(x) + c$, for an arbitrary constant c. We can also change the value of g at its points of discontinuity without changing the value of the integral. We merely note here that a congruence between $BV[a,b]$ and $C^*[a,b]$ can be obtained by introducing the concept of a normalized function of bounded variation and defining an equivalence relation in $BV[a,b]$. Each equivalence class will then contain only one normalized function. Then we will find that $C^*[a,b]$ is congruent to the subspace of $BV[a,b]$ consisting of all normalized functions of bounded variation. For details, see [T3, pp. 197–200].

§6.4. BV_{loc} and L^1_{loc}

We now consider functions defined on unbounded intervals. The results of this section and the corresponding exercises are from [AB].

Definition 6.4.1. Let $f: \mathbf{R}^1 \to \mathbf{R}^1$. If

$$T_f[a,b] < \infty$$

for all $a < b$, then f is said to be *locally of bounded variation* ($f \in BV_{loc}$). If, in addition,

$$\sup_{a<b} T_f[a,b] < \infty,$$

then f is said to be *of bounded variation on $\mathbf{R}^1$* [$f \in BV(\mathbf{R}^1)$] and

$$\sup_{a<b} T_f[a,b] = T_f(-\infty, +\infty).$$

Similar definitions apply to the half-open intervals $[a, +\infty)$ and $(-\infty, b]$.

Definition 6.4.2. Let $f \in BV_{loc}$. The *total variation function* of f is the increasing function defined by

$$Tf(x) = \begin{cases} f(0) + \sup\{\sum |f(x_k) - f(x_{k-1})|\}, & 0 \le x_0 < \cdots < x_n \le x, x \ge 0, \\ f(0) - \sup\{\sum |f(x_k) - f(x_{k-1})|\}, & x \le x_0 < \cdots < x_n \le 0, x \le 0. \end{cases}$$

Clearly,

$$Tf(x) = \begin{cases} f(0) + T_f[0, x], & x \ge 0 \\ f(0) - T_f[x, 0], & x \le 0. \end{cases}$$

Theorem 6.4.3. *For $a \le b$,*

$$Tf(b) - Tf(a) = T_f[a, b].$$

Theorem 6.4.4. *If $f: \mathbf{R}^1 \to \mathbf{R}^1$ is monotonic, then $f \in BV_{loc}$ and*

$$T_f[a, b] = |f(b) - f(a)|.$$

Theorem 6.4.5. *If $f \in BV_{loc}$, let*

$$f_1(x) = \tfrac{1}{2}[Tf(x) + f(x)],$$
$$f_2(x) = \tfrac{1}{2}[Tf(x) - f(x)].$$

Then

$$f(x) = f_1(x) - f_2(x);$$
$$Tf(x) = f_1(x) + f_2(x),$$
$$Tf(0) = f_1(0) = f(0);$$
$$f_2(0) = 0.$$

The functions f_1 and f_2 are called the *positive and negative variations*, respectively, of the function f. (Compare to the functions p and n of Theorem 6.1.15.)

Remark. Theorem 6.4.5 tells us that BV_{loc} can be described as the set of functions which are the difference of two increasing functions. We see that BV_{loc} is also a linear space. Theorem 6.4.5 gives one decomposition of $f \in BV_{loc}$ as the difference of two increasing functions. If, in addition, there are two increasing functions g_1 and g_2 such that $f = g_1 - g_2$ and $g_2(0) = 0$, then

$$0 \le f_i(b) - f_i(a) \le g_i(b) - g_i(a)$$

for $i = 1, 2$ and for all $a \le b$.

Theorem 6.4.6. *Let $f \in BV_{loc}$. Then $[Tf]' = |f'|$ a.e.*

Proof. Consider any closed bounded interval $[a, b]$. For each positive integer n, choose a partition

$$a = x_0^n < \cdots < x_{s(n)}^n = b$$

such that

$$\sum_{k=1}^{s(n)} |\Delta f_k|$$

differs from $Tf(b) - Tf(a)$ by at most 2^{-n}. Define f_n by induction on k as follows:

$$f_n(x) = \begin{cases} Tf(a), & x = a \\ f(x) - f(x_{k-1}^n) + f_n(x_{k-1}^n), & x_{k-1}^n \le x \le x_k^n \text{ and } f(x_k^n) \ge f(x_{k-1}^n) \\ -f(x) + f(x_{k-1}^n) + f_n(x_{k-1}^n), & x_{k-1}^n \le x \le x_k^n \text{ and } f(x_k^n) < f(x_{k-1}^n). \end{cases}$$

Note that on $[x_{k-1}^n, x_k^n]$,

$$f_n(x) = \begin{cases} f(x) + \text{constant}, & f(x_k^n) \ge f(x_{k-1}^n) \\ -f(x) + \text{constant}, & f(x_k^n) < f(x_{k-1}^n), \end{cases}$$

where the constants are chosen so that f_n is defined at x_n^k and $f_n(a) = Tf(a)$. Also note $f_n'(x) = \pm f'(x)$ a.e. and

$$f_n(x_k^n) - f_n(x_{k-1}^n) = |f(x_k^n) - f(x_{k-1}^n)|.$$

Hence,

$$Tf(b) - f_n(b) = Tf(b) - Tf(a) - \sum_{k=1}^{s(n)} |f_n(x_k^n) - f_n(x_{k-1}^n)| \le 2^{-n}.$$

We can then show that $Tf - f$ is increasing on $[a, b]$ and

$$Tf(x) - Tf(x') \ge |f(x) - f(x')| \ge f_n(x) - f_n(x')$$

if $x_{k-1}^n \le x' < x \le x_k^n$. Thus,

$$0 \le \sum_{n=1}^{\infty} \{Tf(x) - f_n(x)\} \le \sum_{n=1}^{\infty} \{Tf(b) - f_n(b)\} \le \sum_{n=1}^{\infty} 2^{-n}.$$

Then we apply Fubini's theorem (1.4.1) to show that $[(Tf)'(x) - f_n'(x)] \to 0$ a.e. Since $f_n'(x) = \pm f'(x)$ a.e. and $(Tf)'(x) \geq 0$ a.e., we have the desired result. □

Remark. Similarly, we can show that if $f \in BV[a,b]$, then for almost all $x \in [a,b]$,

$$|f'(x)| = T_f'[a,x].$$

Theorem 6.4.7. *If $f \in BV_{\text{loc}}$, then f is measurable and integrable on every bounded interval.*

Proof. From Exercise 1, if $f \in BV_{\text{loc}}$, then

$$f = g + s,$$

where $g \in BV_{\text{loc}} \cap C$ and s is the saltus function, which is increasing. Thus, f is measurable. Since f is bounded on every bounded interval, then f is integrable on every bounded interval. □

Theorem 6.4.8. *If $f \in BV_{\text{loc}}$, then f' is measurable.*

Proof. Since f is the difference of two increasing functions, f' is defined. Let

$$f_n(x) = \frac{f(x + 1/n) - f(x)}{1/n}.$$

By Theorem 6.4.7, f is measurable; hence, f_n is measurable. If f is differentiable at x, then

$$\lim_{n \to \infty} f_n(x) = f'(x).$$

Thus,

$$f'(x) = \lim_{n \to \infty} f_n(x)$$

a.e., which implies that f' is measurable. □

Definition 6.4.9. A measurable function g is called *locally integrable* if for every interval $[a,b]$, $g\chi_{[a,b]}$ is integrable, where $\chi_{[a,b]}$ is the characteristic function of the set $[a,b]$. The space of all such locally integrable functions is denoted L^1_{loc}.

Theorem 6.4.10. *If $f \in BV_{\text{loc}}$, then $f' \in L^1_{\text{loc}}$.*

Proof. Since f is the difference of two increasing functions, it suffices to prove the result for f increasing, which we have done in Theorem 1.3.1, showing that

$$\int_a^b f'(x)\,dx \le f(b) - f(a). \qquad\qquad\qquad \Box$$

Remark. Note that Theorem 6.4.10 allows us to talk of a primitive

$$F(t) = \int_0^t f'(x)\,dx$$

if $f \in BV_{\text{loc}}$. But F need not be related to f. For example, if f is a saltus-type function, then $f'(x) = 0$ a.e. (Exercise 1). Then $F \equiv 0$ even though f is not necessarily a constant. We cannot conjecture that $F = f + s$ where s is a saltus-type function, because even this is not true as demonstrated by the Cantor ternary function of Chapter 1. We observe that if $f \in BV_{\text{loc}}$, then

$$h(t) = \int_0^t f'(x)\,dx + c$$

is defined, but the Fundamental Theorem of Calculus does not necessarily hold; i.e., c cannot always be chosen so that $h = f$ a.e. However, we will show that the Fundamental Theorem does hold if $f' \in L^1_{\text{loc}}$.

Theorem 6.4.11. *Let* $g \in L^1_{\text{loc}}$. *If*

$$f(t) = \int_0^t g(x)\,dx + C$$

then $f \in BV_{\text{loc}}$ *and*

$$T_f[a,b] = \int_a^b |g(x)|\,dx.$$

Proof. If $a = x_0 < x_1 \cdots < x_n = b$ is any partition of $[a,b]$, then

$$\sum_{k=1}^n |f(x_k) - f(x_{k-1})| = \sum_{k=1}^n \left| \int_{x_{k-1}}^{x_k} g(x)\,dx \right|$$

$$\le \sum_{k=1}^n \int_{x_{k-1}}^{x_k} |g(x)|\,dx = \int_a^b |g(x)|\,dx.$$

Hence,

$$T_f[a,b] \le \int_a^b |g(x)|\,dx.$$

Now we establish the reverse inequality. Let η be a step function such that $\eta(x) = 0$ if $x \notin [a,b]$ and η assumes the values 1, -1, and 0 only. Then

$$\int_a^b \eta(x)g(x)\,dx \le \sum_{k=1}^n \left| \int_{x_{k-1}}^{x_k} g(x)\,dx \right| = \sum_{k=1}^n |f(x_k) - f(x_{k-1})| \le T_f[a,b].$$

Let $\{\phi_n\}$ be a sequence of step functions that converge to $g\chi_{[a,b]}$ a.e. Assume $\phi_n(x) = 0$ if $x \notin [a,b]$. Let η_n be defined by

$$\eta_n(x) = \begin{cases} 1, & \phi_n(x) > 0 \\ 0, & \phi_n(x) = 0 \\ -1, & \phi_n(x) < 0. \end{cases}$$

Then by the above,

$$\int_a^b \eta_n(x)g(x)\,dx \leq T_f[a,b].$$

But

$$\lim_{n\to\infty} \eta_n(x)g(x) = |g(x)|\chi_{[a,b]}$$

a.e. and

$$|\eta_n g| \leq |g|\chi_{[a,b]}$$

Thus, by the Lebesgue Dominated Convergence Theorem, we have

$$\int_a^b |g(x)|\,dx = \lim_{n\to\infty} \int_a^b \eta_n(x)g(x)\,dx \leq T_f[a,b]. \qquad \square$$

Theorem 6.4.12. *Let* $g \in L^1_{loc}$ *and*

$$f(t) = \int_0^t g(x)\,dx + C.$$

Then $f' = g$ *a.e.*

Proof. Without loss of generality, assume $C = 0$. First, we prove the result for $g = \chi_{[a,b]}$. Then

$$f(t) = \int_0^t g(x)\,dx = \begin{cases} \displaystyle\int_0^a \chi_{[a,b]}(x)\,dx, & t \leq a \\[2mm] \displaystyle t - a + \int_0^a \chi_{[a,b]}(x)\,dx, & a \leq t \leq b \\[2mm] \displaystyle\int_0^b \chi_{[a,b]}(x)\,dx, & b \leq t. \end{cases}$$

Then $f'(t) = g(t)$ for $t \neq a, b$. Since every step function is a linear combination of characteristic functions of intervals, we have $f' = g$ a.e. whenever g is a step function.

Now let N be a positive integer. Since $g \in L^1_{loc}$, g is integrable on $(-N, N)$. Thus, there exists a series $\sum_n \phi_n$ of step functions that is convergent a.e. to $g\chi_{(-N,N)}$ and

$$\sum_n \left|\int \phi_n\right| < +\infty.$$

Hence,

$$\sum_n \int \phi_n^+ < +\infty$$

and

$$\sum_n \int \phi_n^- < +\infty.$$

Let g_1 and g_2 be integrable functions such that

$$g_1 = \sum_n \phi_n^+$$

a.e. and

$$g_2 = \sum_n \phi_n^-$$

a.e. Then

$$g\chi_{(-N,N)} = g_1 - g_2$$

a.e. Let

$$f_1(t) = \int_0^t g_1 = \sum_n \int_0^t \phi_n^+,$$

$$f_2(t) = \int_0^t g_2 = \sum_n \int_0^t \phi_n^-.$$

These are series of increasing functions. Applying Fubini's theorem and note that the derivative of a primitive of ϕ equals ϕ a.e. if ϕ is a step function. Thus,

$$f_1' = \sum_n \phi_n^+$$

a.e. and

$$f_2' = \sum_n \phi_n^-$$

a.e. Hence, for almost all $t \in (-N, N)$,

$$f'(t) = f_1'(t) - f_2'(t) = \sum_n \phi_n^+(t) - \sum_n \phi_n^-(t) = \sum_n \phi_n(t) = g\chi_{(-N,N)}(t). \qquad \square$$

Theorem 6.4.12, along with Exercises 7 and 8, shows that if $f \in BV_{\text{loc}}$ and the Fundamental Theorem of Calculus holds, then

$$f(t) = \int_0^t g(x)\,dx + C,$$

where $g \in L^1_{\text{loc}}$. But it is difficult to check if f is of this form. So we seek another characterization of f which leads us to consider absolute continuity in the next chapter.

Theorem 6.4.13. *Given $E \subset \mathbf{R}^1$ such $mE = 0$, there exists a continuous function f which is of bounded variation such that E is a subset of the points where f is not differentiable.*

Proof. Let $\varepsilon > 0$. Since $mE = 0$, for every integer n, there exists a set $S_n = \{(\alpha_{nk}, \beta_{nk})\}$, a finite or countable collection of open intervals, such that

$$E \subset \bigcup_k (a_{nk}, b_{nk})$$

and

$$\sum_k (\beta_{nk} - \alpha_{nk}) < \frac{\varepsilon}{2^{n+1}}.$$

Let

$$S = \bigcup_{n=1}^{\infty} S_n.$$

Then, renaming the intervals, $S = \{(a_n, b_n)\}$, a countable collection of open intervals such that every $x \in E$ belongs to infinitely many elements of S and

$$\sum_{n=1}^{\infty} (b_n - a_n) < \varepsilon.$$

Now let

$$f_n(x) = \chi_{[a_n, b_n]}(x)(x - a_n) + \chi_{(b_n, \infty)}(b_n - a_n)$$

and

$$f = \sum_{n=1}^{\infty} f_n.$$

Then f is monotone, continuous, and E is a subset of the points of non-differentiability of f. For if $c \in E$ and k is an integer, there exist intervals $(a_{n_1}, b_{n_1}), (a_{n_2}, b_{n_2}), \ldots, (a_{n_k}, b_{n_k})$ in S such that $c \in (a_{n_i}, b_{n_i})$ for each i. Let

$$(a, b) = \bigcap_{i=1}^{k} (a_{n_i}, b_{n_i}).$$

Then for $x \in (a, b)$,

$$\frac{f(x) - f(c)}{x - c} \geq \sum_{i=1}^{k} \frac{f_{n_i}(x) - f_{n_i}(c)}{x - c} \geq k. \qquad \square$$

§6.5. Additional Remarks on Fubini's Theorem

From elementary analysis, we have the result:

If $\sum_{n=1}^{\infty} f_n$ is a series of C^1 functions that is convergent to a function $f \in C^1$ and if $\sum_{n=1}^{\infty} f_n'$ is uniformly convergent on every bounded interval, then $f' = \sum_{n=1}^{\infty} f_n'$.

Although the conclusion of Fubini's Theorem is the same, the hypothesis of his theorem only requires the functions f_n to be increasing; nothing is required of the series $\sum_{n=1}^{\infty} f_n'$.

Theorem 6.5.1 [AB, p. 281]. *Let $\sum_{n=1}^{\infty} f_n$ be a series of BV_{loc} functions. Further, let the series $\sum_{n=1}^{\infty} Tf_n$ of total variation functions be convergent. Then $\sum_{n=1}^{\infty} f_n$ is convergent to a function $f \in BV_{loc}$ and $f' = \sum_{n=1}^{\infty} f_n'$ a.e.*

Proof. Let f_n^1 and f_n^2 be the positive and negative variations, respectively, of f_n. Then

$$Tf_n = f_n^1 + f_n^2,$$
$$f_n = f_n^1 - f_n^2,$$
$$Tf_n(0) = f_n(0) = f_n^1(0),$$
$$f_n^2(0) = 0.$$

By hypothesis, $\sum Tf_n(0) = \sum f_n(0) = \sum f_n^1(0)$ is convergent. For $x \geq 0$,

$$Tf_n(x) - Tf_n(0) = [f_n^1(x) - f_n^1(0)] + [f_n^2(x) - f_n^2(0)]$$
$$\geq f_n^j(x) - f_n^j(0) \geq 0, \quad j = 1, 2.$$

Also,

$$\sum [Tf_n(x) - Tf_n(0)] = \sum Tf_n(x) - \sum Tf_n(0)$$

is convergent; thus, $\sum [f_n^j(x) - f_n^j(0)]$ is convergent. Since $f_n^2(0) = 0$, $\sum f_n^2(x)$ is convergent. Also, $\sum f_n^1(0)$ is convergent, which implies $\sum f_n^1(x)$ is convergent.

Similarly, $\sum f_n^1(x)$ and $\sum f_n^2(x)$ are convergent for $x \leq 0$. Let

$$f^j(x) = \sum f_n^j(x), \quad j = 1, 2.$$

Then f^1 and f^2 are increasing and

$$f^1 - f^2 = \sum [f_n^1 - f_n^2] = \sum f_n.$$

Therefore, $\sum f_n$ converges to the function $f^1 - f^2 = f$ which is locally of bounded variation. Now applying Fubini, we get

$$f' = f^{1'} - f^{2'} = \sum [f_n^{1'} - f_n^{2'}] = \sum [f_n^1 - f_n^2]' = \sum f_n'$$

a.e.   □

EXERCISES

1. Use Fubini's theorem to show that if $f \in BV_{loc}$ and if s is its saltus function, then $s' = 0$ a.e.

2. Let $f_0 \in BV_{loc}$. Let f_1 and f_2 be its positive and negative variations, respectively. Prove that

$$|f_i(x)| \leq |Tf(x)| + 2|f_0(0)|,$$

for $i = 0, 1, 2$.

3. Let $\sum_n f_n = f$ where $f, f_n \in BV_{loc}$. Prove

$$T_f[a,b] \le \sum_n T_{f_n}[a,b].$$

4. If f is bounded and increasing, use Theorem 1.3.1 to show f' is integrable on $\mathbf{R}^1$.

5. [H2, p. 278] Let $t \in (0,1)$. Using induction, define an increasing sequence f_n as follows:

$f_0(x) = x$;

f_n is linear in each interval $\left[\dfrac{k-1}{2^n}, \dfrac{k}{2^n}\right]$, $k = 0, \pm 1, \ldots$;

$f_n(k2^{-n}) = \dfrac{1-t}{2} f_{n-1}\left(\dfrac{k-1}{2^n}\right) + \dfrac{1+t}{2} f_{n-1}\left(\dfrac{k+1}{2^n}\right)$ for k odd;

$f_n\left(\dfrac{k}{2^n}\right) = f_{n-1}\left(\dfrac{k}{2^n}\right)$ for k even.

Prove that

(a) f_n is increasing;
(b) $\{f_n\}$ is increasing and thus converges to a function f;
(c) f is continuous, strictly increasing, and $f' = 0$ a.e.

6. If $m \le x \le m+1$, let $._2 a_1 a_2 \ldots$ be a binary expansion of $x - m$ and let

$$f(x) = m + \sum_{n=1}^{\infty} a_n \left(\frac{1-t}{2}\right)^{p(n)} \left(\frac{1-t}{2}\right)^{n-p(n)},$$

where

$$p(n) = \sum_{k=1}^{n-1} a_k.$$

Prove the series in the definition of f is convergent and independent of the choice of the binary expansion of $x - m$. Show that f is the same function as that of Exercise 5.

7. Prove if $g \in L^1_{loc}$ is continuous at c, then

$$h(t) = \int_0^t g(x)\,dx$$

is differentiable at c and $h'(c) = g(c)$.

8. Prove if $g \in L^1_{loc}$, then

$$f(t) = \int_0^t g(x)\,dx + C$$

is in $BV(\mathbf{R}^1)$ if and only if g is integrable on $\mathbf{R}^1$.

9. Let $E \subset [0, +\infty)$ be a null set. Prove that $\sqrt{E}$ is also a null set.

10. Prove

$$g(x) = \begin{cases} \left| x \sin \dfrac{1}{x} \right|^{1/2}, & x \ne 0 \\ 0, & x = 0 \end{cases}$$

is continuous but not in BV_{loc}.

11. Suppose $\{f_n\}$ is a sequence of functions in BV_{loc} and $f \in BV_{loc}$. To say that f_n converges to f in BV_{loc}, i.e.,

$$f_n - BV_{loc} \to f,$$

means

$$\lim_{n \to \infty} T(f_n - f) = 0$$

everywhere. Prove that if $f_n - BV_{loc} \to f$, then $f_n \to f$ everywhere. [**Hint:** Observe that $Tf_n(0) = f_n(0)$ and then apply Exercise 2 to $f - f_n$.]

12. Suppose $\{f_n\}$ is a Cauchy sequence in BV_{loc}, i.e.,

$$\lim_{n,m \to \infty} T(f_n - f_m) = 0$$

everywhere. Show that there exists $f \in BV_{loc}$ such that $f_n \to f$ everywhere. [**Hint:** Use Exercise 2 to show that, for each x, $\{f_n(x)\}$ is a Cauchy sequence of real numbers. Then use the method of Theorem 6.3.1 to complete the proof.]

13. (Continuation of Exercise 12) Show $f_n - BV_{loc} \to f$.

14. Let $f \in BV(\mathbf{R}^1)$. Define

$$f_1(x) = \tfrac{1}{2}[Tf(x) + f(x)],$$
$$f_2(x) = \tfrac{1}{2}[Tf(x) - f(x)].$$

Show $f = f_1 - f_2$.

15. If $f \in BV(\mathbf{R}^1)$, show

$$Tf(\infty) = \lim_{x \to \infty} Tf(x),$$
$$Tf(-\infty) = \lim_{x \to -\infty} Tf(x),$$
$$f(\infty) = \lim_{x \to \infty} f(x),$$
$$f(-\infty) = \lim_{x \to -\infty} f(x)$$

exist. Prove

$$Tf(\infty) - Tf(-\infty) = \sup_{P \in \mathcal{P}(\mathbf{R}^1)} \sum_k |f(x_k) - f(x_{k-1})|.$$

16. If $f \in BV(\mathbf{R}^1)$, define $\|f\| = Tf(\infty) - Tf(-\infty)$. If $\{f_n\}$ is a sequence in $BV(\mathbf{R}^1)$ and $f \in BV(\mathbf{R}^1)$ such that

$$\lim_{n \to \infty} \|f - f_n\| = 0,$$

show $f_n \to f$ everywhere.

17. If $\{f_n\}$ is a Cauchy sequence in $BV(\mathbf{R}^1)$, show there exists $f \in BV(\mathbf{R}^1)$ such that $f_n \to f$ everywhere.

18. (Continuation of Exercise 17) Show $\lim_{n \to \infty} \|f - f_n\| = 0$.

19. Let $\{f_n\}$ be a sequence in BV_{loc} and $f \in BV_{loc}$ such that

$$\lim_{n \to \infty} T(f - f_n) = 0.$$

Show there exists a subsequence $\{f_{n_j}\}$ such that $f'_{n_j} \to f'$ a.e.

20. Let $f \in C^1$. Prove that $f \in BV_{loc}$ and

$$|Tf(x)| \leq |f(0)| + |x| \sup\{|f'(y)|: -x \leq y \leq x\}.$$

21. Use Exercise 20 to show that if $f_n \in C^1$, $\sum f_n(0)$ converges, and $\sum f'_n$ is uniformly convergent on every bounded interval, then $\sum f_n(x)$ is convergent to $f \in BV_{loc}$ and $f' = \sum f'_n$ a.e.

22. If at every $x \in E$ the derivative $f'(x)$ of a finite function f exists and $|f'(x)| \leq K$, then

$$m^*f(E) \leq K(m^*E).$$

23. Let f be defined and measurable on the set $[a, b]$ and let E be any measurable subset on which f' exists (finitely). Use the previous exercise to show that

$$m^*f(E) \leq \int_E |f'(x)| \, dx.$$

24. Let f be defined on $[a, b]$ and let $E = \{x \in [a, b]: f'(x) = 0\}$. Use Exercise 22 to show that $m[f(E)] = 0$.

25. (a) Show that if k is an integer and

$$f(x) = \begin{cases} x^{2k} \sin x^{-2k+1}, & x \neq 0 \\ 0, & x = 0, \end{cases}$$

then f is differentiable but f' is discontinuous at 0.
(b) Show that if

$$f(x) = \begin{cases} x^2 \sin \dfrac{1}{x^2}, & x \neq 0 \\ 0, & x = 0, \end{cases}$$

f is differentiable but f' is unbounded on $[-1, 1]$.

26. Show that if $h \in BV[a, b]$ and h satisfies the intermediate value property, then h is continuous. [**Hint:** Suppose h is discontinuous at x_0. Assume $d = h(x_0 +) - h(x_0 -) > 0$. Find a $\delta > 0$ such that if $v \in [h(x_0 -) - d/4, h(x_0 +) + d/4] - \{h(x_0)\}$, then h does not assume the value v in $(x_0 - \delta, x_0 + \delta)$, which gives a contradiction.] **Note:** A derivative always satisfies the intermediate value property. Thus, if a function has a derivative which is of bounded variation on a compact interval, then that derivative must be continuous.

Exercise 27–29 yield another proof that if $f \in BV[a, b]$, then f' exists and is finite a.e. These results are found in [A3].

27. If G_1 and G_2 are collections of intervals, then mG_1 will denote the measure of the point set covered by G_1, whereas $G_1 - G_2$ will be the collection of intervals which are in G_1 but not in G_2. Show that any finite collection of intervals G contains a

disjoint subcollection G_1 such that $mG_1 \geq \frac{1}{3}mG$. [**Hint:** Let I_1 be an interval of G of maximal length and $\{I_1\}$ the intervals of G which intersect I_1. Proceed inductively to obtain $I_j \in G - \{\{I_1\}, \ldots, \{I_{j-1}\}\}$ and find k such that $G_1 = \{I_1, \ldots, I_k\}$ has the desired property.]

28. If $f \in BV[a,b]$ and P is any partition of $[a,b]$, let $\pi(x)$ denote the polygonal approximation to f on P and $L(\pi)$ denote the length of the graph of π. Let $f(c,d)$ mean the slope of the chord connecting $(c, f(c))$ with $(d, f(d))$. If $\pi(x)$ is linear on $[a,b]$ with $\pi(a) \leq \pi(b)$ and if $q(x)$ is a polygon coinciding with π at a and b and such that $q(c_i, d_i) < -\alpha$ $(\alpha > 0)$, where $[c_i, d_i]$ is a finite disjoint collection of intervals the sum of whose lengths is d, then

$$L(q) > L(\pi) + d[\sqrt{1 + \alpha^2} - 1].$$

[**Hint:** Translate the sides of q which are parallel to the coordinate axes to obtain an auxiliary polygon q_1 coinciding with q at a and b and with $L(q) = L(q_1)$ and $q_1(a, a + d) < -\alpha$. Let q_2 and q_3 be the two-sided polygons determined by the endpoints of q and $(a + d, q(a + d))$ and $(a + d, q(a))$, respectively.]

29. Use the two preceding exercises to show that if $f \in BV[a, b]$, then Df^+ exists and is finite a.e. [**Hint:** Suppose that f is not differentiable on a set of positive measure. Let E be the set where f is continuous. Then there exist numbers $\alpha > 0$ and β such that if $Df^+ > \beta + \alpha$ and $Df_- < \beta - \alpha$, then $mE > 0$. Without loss of generality, assume $\beta = 0$. Let $\pi(x)$ be any polygonal approximation to f. We may cover each point $x \in E - P$ with an open interval (a_x, b_x) such that π is linear on $[a_x, b_x]$ and $f(a_x, b_x) < -\alpha$ or $>\alpha$ for $\pi(a_x, b_x) \geq 0$ or ≤ 0, respectively. We may pick a finite collection G of the intervals (a_x, b_x) such that $mG > \frac{1}{2}mE$. Use Exercise 27 to get a disjoint subcollection G_1 such that $mG_1 \geq \frac{1}{6}mE$. Now use Exercise 28 to conclude that $L(f) = \infty$, contradicting that $f \in BV$. Use the same type of argument to conclude that f' is finite a.e.

CHAPTER 7

Absolute Continuity

As mentioned in the previous chapter, the problem of reconstructing a function from its derivative leads to the concept of absolute continuity. We begin this chapter with the standard results from introductory texts. Lebesgue points are then introduced. Next we give an extensive collection of sufficient conditions on the derivative to ensure absolute continuity of a function; the hypotheses of these theorems are considerably weaker than those found in introductory texts. Of major importance in establishing absolute continuity is to find conditions on a function or its derivative to ensure that the function possesses the property of mapping null sets into null sets. A counterexample of this property is given in Chapter 8. Emphasis is placed on the relationship between bounded variation and absolute continuity, first by giving conditions to ensure that a function of bounded variation is absolutely continuous; again, the property of mapping null sets into null sets plays an important role. We then introduce singular functions, leading to a decomposition of a function of bounded variation into the sum of three functions, one of which is absolutely continuous. Singular functions will be discussed extensively in Chapter 8. Integration by parts and change of variable results are presented. The chapter concludes with a section on the rectifiability of curves; in this section, we consider vector-valued functions. Rectifiability will be revisited in Chapter 9.

§7.1. Absolute Continuity

In this chapter, we return to the problem of reconstructing the primitive of a function. Specifically, we ask when does the equality

(7.1.1) $$\int_a^x f'(t)\,dt = f(x) - f(a)$$

hold for all x in an interval $[a, b]$. We will see that the answer is if and only if f is absolutely continuous on $[a, b]$. But first we present some examples where (7.1.1) fails.

EXAMPLE 7.1.2 [R5, p. 165]. Let

$$f(x) = \begin{cases} x^2 \sin \dfrac{1}{x^2}, & x \in (0, 1] \\[2mm] 0, & x = 0. \end{cases}$$

Then f is differentiable at every point, but

$$\int_0^1 |f'(t)|\, dt = \infty,$$

so $f' \notin L^1[0, 1]$. Also, $f \notin BV[0, 1]$. If the integral in (7.1.1) (with $[0, 1]$ in place of $[a, b]$) is interpreted as the limit, as $\varepsilon \to 0$, of the integrals over $[\varepsilon, 1]$, then (7.1.1) still holds for this f.

EXAMPLE 7.1.3. The Cantor ternary function and the generalized Cantor function of Exercise 6 of Chapter 1 provide examples of functions which are continuous on $[0, 1]$, differentiable at almost every point of $[0, 1]$ and $f' \in L^1[0, 1]$, but

$$\int_0^1 f'(t)\, dt < f(1) - f(0).$$

Definition 7.1.4. Let f be a finite function defined on the closed interval $[a, b]$. Then f is *absolutely continuous on* $[a, b]$ ($f \in AC[a, b]$) if, for every $\varepsilon > 0$, there exists $\delta > 0$ such that

(7.1.5) $$\sum_{k=1}^n |f(b_k) - f(a_k)| < \varepsilon$$

for any $a \le a_1 < b_1 \le a_2 < b_2 \le \cdots \le a_n < b_n \le b$ for which

$$\sum_{k=1}^n (b_k - a_k) < \delta.$$

Without altering the sense of the definition, we can replace condition (7.1.5) by the stronger condition

(7.1.6) $$\left| \sum_{k=1}^n f(b_k) - f(a_k) \right| < \varepsilon.$$

It is also possible to replace the increments $|f(b_k) - f(a_k)|$ in (7.1.5) by the oscillations of f in the intervals $[a_k, b_k]$.

Remark. When the interval $[a, b]$ is clear, we shall simply write $f \in AC$. As with bounded variation, we also consider those functions which are locally absolutely continuous, i.e., AC_{loc}, and those which are absolutely continuous on the entire real line, i.e., $AC(\mathbf{R}^1)$.

EXAMPLE 7.1.7. A simple example of an absolutely continuous function is one which satisfies a Lipschitz condition on its domain. Thus, a function which has a bounded derivative at every point of its domain is absolutely continuous on its domain.

We state without proof some easy consequences of the definition of absolutely continuity; again the reader may find the details in many texts on real analysis.

Theorem 7.1.8. *If $f \in AC[a,b]$, then f is continuous on $[a,b]$; in fact, it is uniformly continuous.*

Proof. Take $n = 1$ in the definition.   □

Theorem 7.1.9. *If $f \in AC[a,b]$, then $f \in BV[a,b]$.*

Theorem 7.1.10. *If $f, g \in AC[a,b]$, then so are their sum, difference, and product. If $|g(x)| \geq c > 0$ for every $x \in [a,b]$, then the quotient $f/g \in AC[a,b]$. If $c \in \mathbf{R}^1$, then $cf \in AC[a,b]$.*

Remark. That the composition of two AC functions is not necessarily AC is demonstrated in Exercises 3 and 11.

Corollary 7.1.11. *The set of $AC[a,b]$ functions is a linear space.*

Theorem 7.1.12. *If $f \in AC_{loc}$, then the functions Tf, f_1, f_2, i.e., the total variation function and the positive and negative variations, are AC_{loc} functions.*

Theorem 7.1.13. *If $g \in L^1_{loc}$ and*

$$f(t) = \int_0^t g(x)\,dx + c,$$

then $f \in AC_{loc}$. The function f is called an indefinite integral of the function g for every choice of the constant c. "Indefinite" refers to the variable upper limit of integration.

Theorem 7.1.14 [T2, p. 412]. *Let $f \in AC[a,b]$ and let $V(x) = T_f[a,x]$. Then $V \in AC[a,b]$. Thus, f can be expressed as the difference of two increasing absolutely continuous functions.*

Theorem 7.1.15 [T2, p. 413]. *If $f \in AC[a,b]$, then*

$$\int_a^b f'(x)\,dx = f(b) - f(a).$$

Proof. From the previous theorem, we may make the additional assumption that f is increasing. Extend the definition of f and define g_n as in Theorem 1.3.1. Then g_n is continuous and $g_n(x) \to f'(x)$ a.e. As in Theorem 1.3.1, we have

$$\int_a^b g_n(x)\,dx = f(b) - \int_a^{a+1/n} f(x)\,dx,$$

where the integral on the right is a Riemann integral. Because f is continuous,

$$\int_a^b g_n(x)\,dx \to f(b) - f(a).$$

Since we already know that f' is summable, we need only show that

(7.1.16) $$\int_a^b g_n(x)\,dx \to \int_a^b f'(x)\,dx.$$

Let $\varepsilon > 0$. Since $f \in AC[a, b+1]$, there exists $\delta_1 > 0$ such that if $\{(a_i, b_i)\}$ is any collection of disjoint intervals in $[a, b+1]$ satisfying

$$\sum_i (b_i - a_i) < \delta_1,$$

then

$$\sum_i [f(b_i) - f(a_i)] < \frac{\varepsilon}{3}.$$

Also, there is a $\delta_2 > 0$ such that

$$\int_F f'(x)\,dx < \frac{\varepsilon}{3}$$

if $F \subset [a,b]$ and $mF < \delta_2$. We may choose $\delta_2 < \delta_1$. Let E be the set of points in $[a,b]$ where f' exists. By Egoroff's theorem, there is a measurable subset $F \subset E$ with $mF < \delta_2$ such that $g_n \to f'$ uniformly on $E - F$. Then there exists a positive integer N such that

$$\int_{E-F} |g_n(x) - f'(x)|\,dx < \frac{\varepsilon}{3}$$

if $n \geq N$. Thus, we have

$$\int_a^b [g_n(x) - f'(x)]\,dx = \int_E [g_n(x) - f'(x)]\,dx$$

$$= \int_{E-F} [g_n(x) - f'(x)]\,dx + \int_F [g_n(x) - f'(x)]\,dx$$

and, hence,

$$\left| \int_E g_n(x) - \int_E f'(x)\,dx \right| < \frac{\varepsilon}{3} + \frac{\varepsilon}{3} + \int_F g_n(x)\,dx.$$

Thus, to prove (7.1.16), we only need show that, for each n,

(7.1.17)
$$\int_F g_n(x)\,dx < \frac{\varepsilon}{3}.$$

Since $mF < \delta_2 < \delta_1$, there exists an open subset $O \subset (a, b)$ such that $F \subset O$ and $mO < \delta_1$. We can express O as a countable union of disjoint open intervals $\{(a_i, b_i)\}$. For any $x \in [0, 1]$ and any positive integer m, the intervals $[a_i + x, b_i + x], i = 1, 2, \ldots, m$, are all contained in $[a, b + 1]$ and the sum of their lengths is less than δ_1. Hence,

$$\sum_{i=1}^m [f(b_i + x) - f(a_i + x)] < \frac{\varepsilon}{3}.$$

Then

$$\int_{a_1}^{b_1} g_n(x)\,dx = n\left[\int_{a_i}^{b_i} f\left(x + \frac{1}{n}\right)dx - \int_{a_i}^{b_i} f(x)\,dx \right]$$

$$= n\left(\int_{b_i}^{b_i+1/n} f(x)\,dx - \int_{a_i}^{a_i+1/n} f(x)\,dx \right)$$

$$= n \int_0^{1/n} [f(b_i + x) - f(a_i + x)]\,dx.$$

Therefore,

$$\sum_{i=1}^m \int_{a_i}^{b_i} g_n(x)\,dx = n \int_0^{1/n} \left(\sum_{i=1}^m \{f(b_i + x) - f(a_i + x)\} \right)dx \le \frac{\varepsilon}{3}.$$

It follows that

$$\int_F g_n(x)\,dx \le \int_0^b g_n(x)\,dx = \sum \int_{a_i}^{b_i} g_n(x)\,dx < \frac{\varepsilon}{3}. \qquad \square$$

Theorem 7.1.18. Let $f: [a, b] \to \mathbf{R}^1$. Then $f \in AC[a, b]$ if and only if there exists $h \in L^1[a, b]$ such that

$$f(x) = f(a) + \int_a^b h(t)\,dt.$$

It follows that $f' = h$ a.e.

Remark. The result of the previous theorem, i.e., that $h = f'$ a.e., can be improved upon, using the concept of Lebesgue points.

Definition 7.1.19. If

$$\lim_{h \to 0} \frac{1}{h} \int_x^{x+h} |f(t) - f(x)| \, dt = 0$$

at the point x, then x is called a *Lebesgue point* of the function f.

Theorem 7.1.20 [N1, p. 255]. *Let f be summable on $[a, b]$ and define*

$$F(x) = \int_a^x f(t) \, dt.$$

If x is a Lebesgue point of f, then F is differentiable at x and $F'(x) = f(x)$.

Proof. Since

$$\frac{F(x+h) - F(x)}{h} - f(x) = \frac{1}{h} \int_x^{x+h} [f(t) - f(x)] \, dt,$$

we have

$$\left| \frac{F(x+h) - F(x)}{h} - f(x) \right| \le \frac{1}{h} \int_x^{x+h} |f(t) - f(x)| \, dt,$$

which proves the theorem. The converse of this theorem is not true, in general. $\qquad \square$

Theorem 7.1.21 [N1, p. 255]. *If f is summable on $[a, b]$, then almost every point of $[a, b]$ is a Lebesgue point of f.*

Proof. Let r be a rational number. The function $|f(t) - r|$ is summable on $[a, b]$, and hence for almost all $x \in [a, b]$, we have

(7.1.22) $$\lim_{h \to 0} \frac{1}{h} \int_x^{x+h} |f(t) - r| \, dt = |f(x) - r|.$$

Let $E(r)$ be the set of those points in $[a, b]$ for which (7.1.22) does not hold; then $mE(r) = 0$. Let

$$E = \bigcup_{r \in Q} E(r) \cup \{x \in [a, b] : |f(x)| = +\infty\}.$$

Then $mE = 0$. We will show that all points in the set $[a, b] - E$ are Lebesgue points of f. Let $x_0 \in [a, b] - E$ and $\varepsilon > 0$. Let r_n be a rational number such that

$$|f(x_0) - r_n| < \frac{\varepsilon}{3}.$$

Then

$$\left| |f(t) - r_n| - |f(t) - f(x_0)| \right| < \frac{\varepsilon}{3}.$$

Hence,

$$\left| \frac{1}{h} \int_{x_0}^{x_0+h} |f(t) - r_n| \, dt - \frac{1}{h} \int_{x_0}^{x_0+h} |f(t) - f(x_0)| \, dt \right| < \frac{\varepsilon}{3}.$$

Since $x_0 \notin E$, we have

$$\left| \frac{1}{h} \int_{x_0}^{x_0+h} |f(t) - r_n| \, dt - |f(x_0) - r_n| \right| < \frac{\varepsilon}{3},$$

for $|h| < \delta$, i.e.,

$$\frac{1}{h} \int_{x_0}^{x_0+h} |f(t) - r_n| \, dt < \frac{2\varepsilon}{3}.$$

Therefore, for $h < \delta$,

$$\frac{1}{h} \int_{x_0}^{x_0+h} |f(t) - f(x_0)| \, dt < \varepsilon. \qquad \square$$

Theorem 7.1.23 [N1, p. 256]. *Let f be summable on $[a,b]$. Every point of continuity of f is a Lebesgue point of f.*

Proof. Suppose f is continuous at x_0. Then for every $\varepsilon > 0$, there exists $\delta > 0$ such that

$$|f(t) - f(x_0)| < \varepsilon$$

if $|t - x_0| < \delta$. For $|h| < \delta$, we have

$$\frac{1}{h} \int_{x_0}^{x_0+h} |f(t) - f(x_0)| \, dt < \varepsilon,$$

and the result follows. $\qquad \square$

Theorem 7.1.24. *If $f \in AC[a, b]$ and $f' = 0$ a.e, then f is a constant function.*

Proof. Let $\varepsilon > 0$ and

$$E = \{x : x \in (a, b) : f'(x) = 0\}.$$

If $x \in E$, then for sufficiently small $h > 0$,

$$(7.1.25) \qquad \frac{|f(x + h) - f(x)|}{h} < h.$$

The collection of closed intervals $([x, x + h]\}$ for $x \in E$ which satisfy (7.1.25) covers E in the sense of Vitali. Hence, given $\delta > 0$, we can select a finite collection of pairwise disjoint intervals $\{I_k\}$, $k = 1, 2, \ldots, n$, from this collection such that

$$I_k = [x_k, x_k + h_k]$$

and

$$m^*\left(E - \bigcup_{k=1}^{n} I_k\right) < \delta.$$

We can order the I_k such that $x_k < x_{k+1}$. If

$$[a, x_1), (x_1 + h_1, x_2), \ldots, (x_{n-1} + h_{n-1}, x_n), (x_n + h_n, b]$$

are the intervals remaining after removing the I_k, then the sum of their lengths must be less than δ since

$$b - a = mE \le \sum_{k=1}^{n} mI_k + m^*\left(E - \bigcup_{k=1}^{n} I_k\right) < \sum_{k=1}^{n} mI_k + \delta.$$

Thus,

$$\sum_{k=1}^{n} mI_k > b - a - \delta.$$

Since $f \in AC$, δ can be chosen small enough that

$$(7.1.26) \quad \left| f(x_1) - f(a) + \sum_{k=2}^{n} \{f(x_k) - f(x_{k-1} + h_{k-1})\} + f(b) - f(x_n + h_n) \right| < \varepsilon.$$

By definition of the interval I_k,

$$|f(x_k + h_k) - f(x_k)| < \varepsilon h_k.$$

Then

$$(7.1.27) \quad \left| \sum_{k=1}^{n} f(x_k + h_k) - f(x_k) \right| < \varepsilon(b - a)$$

since

$$\sum_{k=1}^{n} h_k = \sum_{k=1}^{n} mI_k \le b - a.$$

Combining (7.1.26) and (7.1.27) yields

$$|f(b) - f(a)| < \varepsilon(1 + b - a).$$

Since ε is arbitrary, we must have $f(a) = f(b)$. The same reasoning can be applied to every interval $[a, x]$ for $x \in (a, b)$ to get $f(x) = f(a)$. Therefore, f is constant on $[a, b]$. $\square$

We now present some theorems which give sufficient conditions on the derivative of a function to ensure the absolute continuity of that function.

Theorem 7.1.28. *If f is differentiable on $[a, b]$ and f' is bounded on $[a, b]$, then $f \in AC$ and, hence,*

$$f(t) = \int_{0}^{t} f'(x) \, dx + f(a).$$

Proof. Use the mean value theorem. In elementary calculus, we prove this result under the additional hypothesis that f' is continuous to assure that the derivative of

$$\int_0^t f'(x)\,dx$$

is f' everywhere. □

Lemma 7.1.29 [G5]. *If f has a finite derivative at every point of an interval $[a, b]$, then f maps null sets into null sets.*

Proof. Let $S \subset [a, b]$ have measure zero. For each positive integer n, let

$$S_n = \{x \in S : |f'(x)| \le n\}.$$

Then $mS_n = 0$ and by Exercise 22 of Chapter 6

$$m^*(f(S_n)) \le n \cdot m^*(S_n) = 0.$$

Thus, $mf(S_n) = 0$. Since

$$S = \bigcup_{n=1}^{\infty} S_n$$

and

$$f(S) = f\left(\bigcup_{n=1}^{\infty} S_n\right) = \bigcup_{n=1}^{\infty} f(S_n),$$

we have

$$m^*(f(S)) \le \sum_{n=1}^{\infty} m^*(f(S_n)) = 0.$$

Thus, $mf(S) = 0$. □

Remark. An example of a continuous function with a finite derivative a.e. with maps a null set into a set of positive measure is given in Statement 8.1.18.

Corollary 7.1.30 [G5]. *If f has a finite derivative at every point of an interval $[a, b]$ and $\{I_n\}$ is a sequence of nonoverlapping intervals in $[a, b]$ with*

$$\sum_{n=1}^{\infty} mI_n = mI,$$

then

$$\sum_{n=1}^{\infty} m[f(I_n)] \ge |f(b) - f(a)|.$$

Proof. Since $I_n \subset I$ for each n and $\sum_{n=1}^{\infty} mI_n = mI$, we have

$$I = \bigcup_{n=1}^{\infty} I_n \cup N,$$

where N is a set of measure zero. Hence,

$$|f(b) - f(a)| \leq mf(I) = \sum_{n=1}^{\infty} mf(I_n) + mf(N).$$

By the previous lemma, $f(N)$ has measure zero, and the result follows.   □

Lemma 7.1.31. [G5]. *If f has a finite derivative at every point of the interval $[a,b]$ and f' is summable on $[a,b]$, then for every $[c,d] \subset [a,b]$,*

$$\int_c^d |f'(t)|\, dt \geq f(d) - f(c).$$

Proof. Because f' is summable on $[a,b]$, almost every $x \in [a,b]$ is a Lebesgue point of f', i.e.,

$$\lim_{h \to 0} \frac{1}{h} \int_x^{x+h} |f'(t) - f(x)|\, dt = 0$$

for almost all $x \in [a,b]$ and hence for almost $x \in [c,d]$. Since

$$\left| \left| \frac{1}{h} \int_x^{x+h} f'(t)\, dt \right| - |f'(x)| \right| \leq \left| \frac{1}{h} \int_x^{x+h} \{f'(t) - f'(x)\}\, dt \right|$$

$$\leq \frac{1}{h} \int_x^{x+h} |f'(t) - f'(x)|\, dt,$$

we have

$$\lim_{h \to 0} \left| \frac{1}{h} \int_x^{x+h} f'(t)\, dt \right| = |f'(x)|$$

if x is a Lebesgue point. Now

$$m[f([x, x+h])] \geq |f(x+h) - f(x)|,$$

so that

$$\lim_{h \to 0} \frac{1}{h} m[f([x, x+h])] \geq |f'(x)|.$$

By Exercise 23 of Chapter 6 (with $E = [x, x+h]$),

$$m[f([x, x+h])] \leq \int_x^{x+h} |f'(t)|\, dt.$$

Hence,

$$\frac{1}{h}m[f([x, x + h])] - |f'(x)| \leq \frac{1}{h}\int_x^{x+h} |f'(t)|\, dt - |f'(x)|.$$

But we have just shown above that the right-hand side of the above inequality converges to 0 as $h \to 0$. This implies that

$$\lim_{h\to 0}\frac{1}{h}m[f([x, x + h])] - |f'(x)| \leq 0.$$

Therefore,

$$\lim_{h\to 0}\frac{1}{h}m[f([x, x + h])] = |f'(x)|.$$

Combining these two results, we have, for almost all $x \in [c, d]$,

$$\lim_{h\to 0}\left|\frac{1}{h}\int_x^{x+h} f'(t)\, dt\right| = \lim_{h\to 0}\frac{1}{h}m[f([x, x + h])] = |f'(x)|.$$

Let $\varepsilon > 0$. Then for almost all $x \in [c, d]$, there exists $h_x > 0$ such that if $0 < h < h_x$,

$$\left|\left\|\int_x^{x+h} f'(t)\, dt\right| - m[f([x, x + h])]\right| < \varepsilon h,$$

from which it follows that

$$\int_x^{x+h} |f'(t)|\, dt > m[f([x, x + h])] - \varepsilon h.$$

By the Vitali Covering Theorem, there are pairwise disjoint intervals $[c_n, d_n]$, $n = 1, 2, \ldots$ with

$$\sum_{n=1}^{\infty}(d_n - c_n) = d - c$$

and

$$\int_c^d |f'(t)|\, dt = \sum_{n=1}^{\infty}\int_{c_n}^{d_n} |f'(t)|\, dt$$

$$\geq \sum_{n=1}^{\infty} m[f([c_n, d_n])] - \varepsilon(d - c) \geq |f(d) - f(c)| - \varepsilon(d - c)$$

by the previous corollary. Since ε is arbitrary,

$$\int_c^d |f'(t)|\, dt > |f(d) - f(c)|. \qquad \square$$

Theorem 7.1.32 [G5]. *If f' exists (finitely) everywhere and is summable on an interval $[a, b]$, then f is absolutely continuous on $[a, b]$.*

Proof. For $\varepsilon > 0$, there exists $\delta > 0$ such that if $[a_i, b_i]$, $i = 1, 2, \ldots, n$ are pairwise disjoint intervals and if

$$\sum_{i=1}^{n} (b_i - a_i) < \delta,$$

then

$$\sum_{i=1}^{n} |f(b_i) - f(a_i)| \leq \sum_{i=1}^{n} \int_{a_i}^{b_i} |f'(t)|\, dt < \varepsilon. \qquad \square$$

We now prove that the conclusion of Theorem 7.1.32 holds under some weaker hypotheses. We first consider the case where the derivative exists everywhere and is summable on $[a, b]$; note this implies that the derivative is finite a.e. Then we further weaken the hypothesis to require that the function be continuous, that its derivative exist (finitely) at all points of $[a, b]$ except possible on a set which is finite or enumerable, and that the derivative is summable on $[a, b]$. However, the result does not hold under the condition that the function is continuous, its derivative exists except on a set of measure zero, and the derivative is summable, as illustrated by the Cantor ternary function. However, if the function is continuous, its derivative exists except on a set of measure zero, the derivative is summable, and the function maps null sets into null sets, then the function is absolutely continuous. This last result is left as an exercise.

Lemma 7.1.33 [R5, p. 54]. *For any $f \in L^1[a, b]$ and any $\varepsilon > 0$, there exist functions u and v such that $u \leq f \leq v$, u is upper semicontinuous and bounded from above while v is lower semicontinuous and bounded from below, and*

$$\int_{a}^{b} (v - u) < \varepsilon.$$

Proof. Assume first that $f \geq 0$ and that f is not identically zero. Since f is the pointwise limit of an increasing sequence of simple functions s_n, f is the sum of the simple functions $t_n = s_n - s_{n-1}$ (letting $s_0 = 0$). Because each t_n is a linear combination of characteristic functions, there exist measurable sets E_i (not necessarily disjoint) and constants $c_i > 0$ such that

$$f(x) = \sum_{i=1}^{\infty} c_i \chi_{E_i}(x), \quad x \in [a, b].$$

Since

$$\int_{a}^{b} f(x)\, dx = \sum_{i=1}^{\infty} c_i m E_i,$$

the series on the right converges. We can find compact sets K_i and open sets V_i such that $K_i \subset E_i \subset V_i$ and

$$c_i m(V_i - K_i) < 2^{-i-1}\varepsilon, \quad i = 1, 2, \ldots.$$

Let

$$v = \sum_{i=1}^{\infty} c_i \chi_{V_i}, \qquad u = \sum_{i=1}^{\infty} c_i \chi_{K_i},$$

where N is chosen so that

$$\sum_{i=N+1}^{\infty} c_i m E_i < \frac{\varepsilon}{2}.$$

Then v is lower semicontinuous, u is upper semicontinuous, $u \le f \le v$, and

$$v - u = \sum_{i=1}^{N} c_i(\chi_{V_i} - \chi_{K_i}) + \sum_{i=N+1}^{\infty} c_i \chi_{V_i}$$

$$\le \sum_{i=1}^{\infty} c_i(\chi_{V_i} - \chi_{K_i}) + \sum_{i=N+1}^{\infty} c_i \chi_{E_i}.$$

The result now follows. The general case is proved by writing $f = f^+ - f^-$ and applying the above to each of the functions f^+ and f^-. The result also holds for measurable sets as well as intervals.   □

Theorem 7.1.34 [R5, p. 168]. *Suppose f is differentiable at every point of the interval $[a, b]$ and that $f' \in L^1[a, b]$. Then f is absolutely continuous, i.e.,*

$$f(x) - f(a) = \int_a^x f'(t) \, dt \quad (a \le x \le b).$$

Proof. We will prove the result for the case $x = b$. Let $\varepsilon > 0$. By the previous lemma, there exists a lower semicontinous function g such that $g > f'$ and

(7.1.35) $$\int_a^b g(t) \, dt < \int_a^b f'(t) \, dt + \varepsilon.$$

Actually, the lemma only gives $g \ge f'$, but since $m[a, b] < \infty$, we can increase g by a small constant without affecting (7.1.35). Let $\eta > 0$ and define

(7.1.36) $$F_\eta(x) = \int_a^x g(t) \, dt - f(x) + f(a) + \eta(x - a), \quad a \le x \le b.$$

For each $x \in [a, b)$, there exists $\delta_x > 0$ such that

$$g(t) > f'(x) \qquad \text{and} \qquad \frac{f(t) - f(x)}{t - x} < f'(x) + \eta$$

for all $t \in (x, x + \delta_x)$ since g is lower semicontinuous and $g(x) > f'(x)$. Therefore, for any such t, we have

$$F_\eta(t) - F_\eta(x) = \int_x^t g(s) \, ds - [f(t) - f(x)] + \eta(t - x)$$

$$> (t - x)f'(x) - (t - x)[f'(x) + \eta] + \eta(t - x) = 0.$$

Because $F_\eta(a) = 0$ and F_η is continuous, there is a last point $x \in [a,b]$ at which $F_\eta(x) = 0$. If $x < b$, then the above implies that $F_\eta(t) > 0$ for $t \in (x,b]$. Thus, $F_\eta(b) \geq 0$. Since η was arbitrary, (7.1.35) and (7.1.36) give

$$f(b) - f(a) \leq \int_a^b g(t)\,dt < \int_a^b f'(t)\,dt + \varepsilon.$$

Because ε was arbitrary, we conclude that

$$f(b) - f(a) \leq \int_a^b f'(t)\,dt.$$

Since $-f$ also satisfies the hypotheses of the theorem, the above inequality holds with $-f$ in place of f. The two inequalities together yield the desired conclusion. $\square$

Theorem 7.1.37 [K2, p. 183]. *Suppose f is continuous throughout a closed interval $[a,b]$ and that f' is defined and finite in $[a,b]$ except at points of a set E which is finite or enumerable, and f' is summable on $[a,b]$. Then*

$$\int_a^b f'(x)\,dx = f(b) - f(a).$$

Proof. Let n be any positive integer and

$$g_n(x) = \min(n, \overline{D}f(x))$$

for $x \in (a,b)$. Then $g_n(x)$ is measurable. If $x \in [a,b] \cap E^c$, then $|g_n(x)| \leq |f'(x)|$ and

$$\lim_{n \to \infty} g_n(x) = f'(x).$$

Since E is of measure zero, it follows that

$$\int_a^b f'(x)\,dx = \lim_{n \to \infty} \int_a^b g_n(x)\,dx.$$

Let

$$G_n(x) = \int_a^x g_n(t)\,dt.$$

If $x, x + h \in [a,b]$, then

$$G_n(x + h) - G_n(x) = \int_x^{x+h} g_n(t)\,dt.$$

So if $h \neq 0$, we have

$$\frac{G_n(x + h) - G_n(x)}{h} \leq n.$$

Let $\theta_n(x) = f(x) - G_n(x)$. If $x \in [a,b] \cap E^c$, then

$$\underline{D}\theta_n(x) = f'(x) + \underline{D}\{-G_n(x)\} \geq f'(x) - n > -\infty.$$

Also, since $G_n'(x) = g_n(x)$ a.e. and $mE = 0$, it follows that for almost all $x \in [a,b]$,

$$\underline{D}\theta_n(x) = f'(x) - g_n(x) \geq 0.$$

Thus, we have shown that set $\{x \in [a,b]: \underline{D}\theta_n(x) = -\infty\}$ is finite or enumerable and that the set $\{x \in [a,b]: \underline{D}\theta_n(x) < 0\}$ has measure zero. Since $\theta_n(x)$ is continuous on $[a,b]$, it follows from Theorem 5.2.3 that $\theta_n(b) \geq \theta_n(a)$, i.e.,

$$f(b) - \int_a^b g_n(x) \geq f(a), \quad n = 1, 2, \ldots .$$

Hence,

$$f(b) - f(a) \geq \int_a^b f'(x)\,dx.$$

Since $-f$ also satisfies the hypotheses of the theorem, the above inequality holds with $-f$ in place of f. The two inequalities together yield the desired conclusion. $\qquad \square$

Theorem 7.1.38. *A function $f \in BV_{\text{loc}}$ is in AC_{loc} if and only if*

(a) *f is continuous,*
(b) *f maps null sets into null sets.*

Proof. We have already observed that an absolutely continuous function is continuous. Now suppose $f \in AC_{\text{loc}}$ and E is a set of measure zero. We first consider the case where E is contained in a finite interval $[a,b]$. Since $\{f(a)\}$ and $\{f(b)\}$ are null sets, we can assume $E \subset (a,b)$. Let $\varepsilon > 0$. From the definition of absolute continuity, there exists $\delta > 0$, such that if $\{(a_k, b_k)\}$ is any collection of pairwise disjoint intervals contained in $[a,b]$ for which

$$\sum_k (b_k - a_k) < \delta,$$

then

$$\sum_k |f(b_k) - f(a_k)| < \varepsilon.$$

For this δ, there exists an open set O such that $E \subset O$ and $m(O - E) < \delta$. Thus, $mO < \delta$. We may also assume that $O \subset (a,b)$. Each open set in $\mathbf{R}^1$ is the union of a countable collection of pairwise disjoint open intervals. Let

$$O = \bigcup_{n=1}^{\infty} I_n,$$

where each $I_n = (x_{n-1}, x_n)$ and $I_j \cap I_k$ is empty if $k \neq j$. Since $E \subset O$,

$$f(E) \subset f(O) = f\left(\bigcup_{n=1}^{\infty} I_n\right) = \bigcup_{n=1}^{\infty} f(I_n) \subset \bigcup_{n=1}^{\infty} f(\bar{I}_n),$$

where $\bar{I}_n = [x_{n-1}, x_n]$, the closure of the open interval I_n. Since each $\bar{I}_n$ is a compact interval,

$$f(\bar{I}_n) = [m_n, M_n],$$

where

$$m_n = \min\{f(x): x \in \bar{I}_n\},$$
$$M_n = \max\{f(x): x \in \bar{I}_n\}.$$

Since f is continuous on each $\bar{I}_n$, there exist points $\alpha_n, \beta_n \in \bar{I}_n$ such that

$$f(\alpha_n) = m_n,$$
$$f(\beta_n) = M_n.$$

Then

$$\sum_n |\alpha_n - \beta_n| \leq \sum_n (x_n - x_{n-1}) = \sum_n mI_n = mO < \delta,$$

from which it follows that

$$\sum_n (M_n - m_n) < \varepsilon.$$

Thus,

$$m^*[f(E)] \leq m^*[f(O)] \leq \sum_n (M_n - m_n) < \varepsilon.$$

Since ε was arbitrary, $m[f(E)] = 0$. The general case now follows easily, since for each n

$$E_n = E \cap [-n, n]$$

is a bounded set of measure zero.

Now we prove the converse. Suppose f is an increasing continuous function which maps null sets into null sets and f is not absolutely continuous on the interval $[a, b]$. Then there exists ε_0 such that for every $\delta > 0$, there is a set of pairwise disjoint open intervals $\{(a_k, b_k)\}$ for which

$$\sum_k (b_k - a_k) < \delta$$

and

$$\sum_k (M_k - m_k) \geq \varepsilon_0.$$

Let

$$\sum_{i=1}^{\infty} \delta_i$$

be a convergent series of positive terms. For every δ_i, let $\{(a_k^i, b_k^i)\}$, $k = 1, 2,$ $\ldots, n_i$, be a collection of pairwise disjoint intervals for which

$$\sum_{k=1}^{n_i} (b_n^i - a_n^i) < \delta_i$$

and

$$\sum_{k=1}^{n_i} (M_k^i - m_k^i) \geq \varepsilon_0,$$

where M_k^i and m_k^i denote the maximum and minimum values, respectively, of f in the interval $[a_k^i, b_k^i]$. Let

$$E_i = \bigcup_{k=1}^{n_i} (a_k^i, b_k^i)$$

and

$$A = \bigcap_{n=1}^{\infty} \bigcup_{i=1}^{\infty} E_i.$$

Then $mA = 0$, from which it follows from the hypothesis that $m[f(A)] = 0$.

Now define functions $L_k^i(y)$, $k = 1, 2, \ldots, n_i$, $i = 1, 2, \ldots$, as follows: $L_k^i(y) = 1$ if there is at least one $x \in (a_k^i, b_k^i)$ for which $f(x) = y$; otherwise, $L_k^i(y) = 0$. Clearly, $L_k^i(y) = 1$ for all $y \in (m_k^i, M_k^i)$ and $L_k^i(y) = 0$ for all $y \notin [m_k^i, M_k^i]$. Therefore,

(7.1.39)
$$\int_m^M L_k^i \, dy = M_k^i - m_k^i,$$

where M and m are the maximum and minimum, respectively, of f on the interval $[a, b]$. Let

$$N_i(y) = \sum_{k=1}^{n_i} L_k^i(y).$$

Then $N_i(y)$ is the number of those intervals (a_k^i, b_k^i) containing at least one solution to $f(x) = y$. Hence,

(7.1.40)
$$N_i(y) \leq N(y),$$

where $N(y)$ is the Banach indicatrix of the function f. By (7.1.39),

(7.1.41)
$$\int_m^M N_i(y) \, dy \geq \varepsilon_0.$$

Now we will show that for almost all $y \in [m, M]$,

(7.1.42)
$$\lim_{i \to \infty} N_i(y) = 0.$$

Since the Banach indicatrix is summable, it will follow from (7.1.40) and

(7.1.42) that

$$\lim_{i \to \infty} \int_m^M N_i(y) \, dy = 0,$$

contradicting (7.1.41), proving our assumption that $f \notin AC$ is false. Let B be the set of y for which (7.1.42) does not hold and let C be the set of y for which $N(y) = +\infty$. Since $N(y)$ is summable, $mC = 0$. To complete the proof, we must verify that

(7.1.43) $B - C \subset f(A).$

Let $y_0 \in B - C$. Since the functions $N_i(y)$ assume only non-negative integral values, there is a sequence $\{i_r\}$ of natural numbers such that

$$N_{i_r}(y_0) \geq 1.$$

For every r, there exists a point $x_{i_r} \in E_{i_r}$ such that

$$f(x_{i_r}) = y_0.$$

Since $N(y_0) < +\infty$, there are only a finite number of distinct points among the points x_{i_r}. Hence, one of them, call it x_0, occurs an infinite number of times in the sequence $\{x_{i_r}\}$. The point x_0 belongs to an infinite number of the sets E_i, and, clearly, $f(x_0) = y_0$. Thus, $x_0 \in A$ and $y_0 \in f(A)$. This verifies (7.1.43).   □

Definition 7.1.44. A *singular function* is a nonconstant function whose derivative exists and is equal to zero a.e.

Theorem 7.1.45 [J2, p. 544]. *Let $f: [a, b] \to \mathbf{R}^1$ be increasing. Then f can be expressed in the form*

$$f = F + g + s,$$

where F is increasing and absolutely continuous, g is an increasing singular function, and s is a saltus-type function. In this decompositon, the components F, g, and s are unique to within additive constants.

Proof. Recall from Theorem 1.1.6 that an increasing function can be expressed in the form

$$f(x) = \phi(x) + s(x),$$

where ϕ is an increasing continuous function and s is the saltus function of f. In Exercise 1 of Chapter 6, it was shown that $s' = 0$ a.e. Thus, $f' = \phi'$ a.e. Define

$$F(x) = \int_a^x f'(t) \, dt.$$

Then $F' = f' = \phi'$ a.e. Hence,

$$g(x) = \phi(x) - F(x)$$

is continuous and $g' = 0$ a.e. Also, g is increasing. To see this, suppose $a \leq x < y \leq b$. Then

$$g(x) - g(y) = \phi(x) - \phi(y) - F(y) - F(x)$$

$$= \phi(x) - \phi(y) + \int_x^y \phi'(t)\,dt$$

$$\leq \phi(x) - \phi(y) + \phi(y) - \phi(x)$$

$$= 0.$$

Thus,

$$f(x) = \int_a^x f'(t)\,dt + g(x) + s(x)$$

$$= F(x) + g(x) + s(x).$$

Now we prove the uniqueness of this representation. Suppose

$$f = F_1 + g_1 + s_1$$

is another representation of the same variety. Then $f' = F'$ a.e. and $f' = F_1'$ a.e. Apply Theorem 7.1.24 to the absolutely continuous function $F - F_1$ to conclude $F - F_1$ is a constant. That the remaining quantities $g - g_1$ and $s - s_1$ are constant follows from Exercise 4 of Chapter 1.   □

Theorem 7.1.46 [J2, p. 552]. *Let $f \in BV[a,b]$. Then f can be expressed in the form*

$$f = F + g + s,$$

where F is absolutely continuous, g is a continuous singular function, and s is the difference of two increasing saltus-type functions. The functions F, g, and s are unique to within additive constants. If, in addition, f is continuous, then

$$f = F + g,$$

where F is absolutely continuous and g is a continuous singular function.

Proof. This result follows from the previous theorem and the fact that f is the difference of two increasing functions.   □

Theorem 7.1.47 (Integration by Parts [J2, p. 553]). *If $f, g \in AC[a,b]$, then*

$$\int_a^b f(x)g'(x)\,dx = f(b)g(b) - f(a)g(a) - \int_a^b f'(x)g(x)\,dx.$$

Proof. Since the product fg is absolutely continuous,

$$\int_a^b (fg(x))' \, dx = f(b)g(b) - f(a)g(a).$$

Now apply the product rule to $(fg)'$. $\square$

Theorem 7.1.48 (Change of Variable [J2, p. 560]). *Let h be an increasing absolutely continuous function defined on $[a, b]$. If f is integrable on $[h(a), h(b)]$, then $(f \circ h) \cdot h'$ is integrable on $[a, b]$ and*

$$\int_{h(a)}^{h(b)} f(y) \, dy = \int_a^b (f \circ h) \cdot h'(x) \, dx.$$

Proof. We first suppose that f is the characteristic function of an open interval $(c, d) \subset (h(a), h(b))$. Choose the largest α and the smallest β in $[a, b]$ such that $h(\alpha) = c$ and $h(\beta) = d$. Thus, $c < h(x) < d$ if and only if $\alpha < x < \beta$. Therefore,

$$\int_a^b f(h(x))h'(x) \, dx = \int_\alpha^\beta h'(x) \, dx$$

$$= h(\beta) - h(\alpha)$$

$$= d - c$$

$$= \int_{h(a)}^{h(b)} f(y) \, dy.$$

Since every open set is the countable union of disjoint open intervals, the result is valid if $f = \chi_G$ for any open set G. By taking complements, it can be seen to be valid for $f = \chi_K$ where K is any compact set.

Now let $E \subset [h(a), h(b)]$ be any measurable set. There exist compact sets K_j and open sets G_j such that

$$K_1 \subset K_2 \subset \cdots \subset E \subset \cdots \subset G_2 \subset G_1$$

and

$$m(G_j - K_j) \to 0.$$

Let

$$\phi_j(x) = \chi_{K_j}(h(x))h'(x),$$

$$\psi_j(x) = \chi_{G_j}(h(x))h'(x),$$

$$g(x) = \chi_E(h(x))h'(x).$$

Thus,

$$\phi_1 \leq \phi_2 \leq \cdots \leq g \leq \cdots \leq \psi_2 \leq \psi_1.$$

Then ϕ_j and ψ_j are measurable, and

$$\int_a^b \phi_j(x)\,dx = \int_{h(a)}^{h(b)} \chi_{K_j}(y)\,dy = mK_j,$$

$$\int_a^b \psi_j(x)\,dx = \int_{h(a)}^{h(b)} \chi_{G_j}(y)\,dy = m(G_j \cap [h(a), h(b)]).$$

Thus,

$$\int_a^b \lim \phi_j(x)\,dx = mE = \int_a^b \lim \psi_j(x)\,dx.$$

Since

$$\lim \phi_j \leq g \leq \lim \psi_j,$$

this proves that g is measurable and

$$\int_a^b g(x)\,dx = mE.$$

Thus, we have proved the theorem when $f = \chi_E$, where E is any measurable set. The result may now be extended to any $f \in L^1(h(a), h(b))$. □

We can apply our result that absolutely continuous functions are also of bounded variation and, hence, are the difference of two increasing functions to obtain the following generalization of Theorem 7.1.48.

Theorem 7.1.49. *Let $h \in AC[a, b]$ and f a bounded integrable function. Then $(f \circ h) \cdot h' \in L^1(a, b)$ and*

$$\int_{h(a)}^{h(b)} f(x)\,dx = \int_a^b (f \circ h) \cdot h'(y)\,dy.$$

§7.2. Rectifiable Curves

A concept closely related to bounded variation is that of the length of a curve. In this section, we shall allow functions to be vector-valued. Thus, we shall consider functions of the form

$$f: [a, b] \to \mathbf{R}^n$$

and we shall use the Euclidean norm. If $v = (v_1, \ldots, v_n) \in \mathbf{R}^n$, then the *norm* of v is

$$\|v\| = (v_1^2 + \cdots + v_n^2)^{1/2}.$$

The function f has n component functions $f_1, \ldots, f_n$, each of whose range is

$\mathbf{R}^1$, i.e.,

$$f(x) = (f_1(x), \ldots, f_n(x)).$$

The total variation of a vector-valued function is defined in the same way as the total variation of a real-valued function, simply replace absolute value by norm in Definition 6.1.2.

Definition 7.2.1. A *curve* in $\mathbf{R}^n$ is a function $f: [a, b] \to \mathbf{R}^n$. The interval $[a, b]$ is called the parameter interval for the curve f and f is called a *parameterized* curve. We shall use the letter t to denote the parameter, $a \le t \le b$. In this section we shall assume that a curve is continuous.

Definition 7.2.2. For any partition

$$P = \{a = t_0, t_1, \ldots, t_m = b\}$$

of $[a, b]$, the points $f(t_0)$, $f(t_1)$, $\ldots$, $f(t_m)$ are the vertices of an inscribed polygon. The length of the polygon is

$$\sum_{k=1}^{m} \|f(t_k) - f(t_{k-1})\|.$$

We define the *length* of the curve f to be the least upper bound of the lengths of such polygons, i.e, the length of the curve is just the total variation $T_f[a, b]$. If this length is finite, we say the curve is *rectifiable*. Thus, f is rectifiable if and only if $f \in BV$.

Theorem 7.2.3. *Let $f: [a, b] \to \mathbf{R}^n$ be a curve with components $f = (f_1, \ldots, f_n)$. Then f is rectifiable if and only if each component function is in $BV[a, b]$.*

Proof. The result follows immediately from the following inequality: If $\mathbf{x} = (x_1, \ldots, x_n) \in \mathbf{R}^n$, then

$$|x_k| \le \|\mathbf{x}\| \le |x_1| + \cdots + |x_n|.$$

Thus,

$$T_{f_k}[a, b] \le T_f[a, b] \le T_{f_1}[a, b] + \cdots + T_{f_n}[a, b]. \qquad \square$$

Remark. With the previous theorem, many of our results on real-valued functions become valid for vector-valued functions, among them 6.1.5, 6.1.6, 6.1.7, 6.1.9, 6.1.10, 6.1.11, 6.1.12, 6.1.19, 6.3.6, 6.3.11, and 6.3.12.

Definition 7.2.4. Let $f: [a, b] \to \mathbf{R}^n$ be a rectifiable curve. The *arc length function* of f is the function $s: [a, b] \to \mathbf{R}^1$ defined by

$$s(t) = T_f[a, t].$$

Summarizing the results we have obtained on bounded variation and absolute continuity, we have the following:

(7.2.5) s is increasing and continuous;

(7.2.6) if $t < t'$, $T_f(t, t') = s(t') - s(t)$;

(7.2.7) $\|f'(t)\| = s'(t)$ a.e.;

(7.2.8) s is absolutely continuous if and only if each component function of f is absolutely continuous;

(7.2.9) If L is the length of f,

$$\int_a^b s'(t)\,dt = \int_a^b \|f'(t)\|\,dt \le L,$$

with equality if and only f is absolutely continuous.

Theorem 7.2.10 [J2, p. 565]. *Let $f: [a, b] \to \mathbf{R}^n$ be a rectifiable curve. Then for any $\varepsilon > 0$ there exists $\delta > 0$ such if $P = \{a = t_0, t_1, \ldots, t_m = b\}$ is a partition of $[a, b]$ and $\|P\| < \delta$, then*

$$\sum_{k=1}^m \|f(t_k) - f(t_{k-1})\| > T_f[a, b] - \varepsilon.$$

Proof. We remark that the essence of this result is that when f is a *continuous* function of bounded variation, then any sufficiently fine partition generates a good approximation of the total variation and, hence, of the length of the curve.

Let $\varepsilon > 0$. First, we choose a partition $P_0 = \{a = T_0, T_1, \ldots, T_M = b\}$ such that

$$V_0 = \sum_{k=1}^M \|f(T_k) - f(T_{k-1})\| > T_f[a, b] - \frac{\varepsilon}{2}.$$

Since f is uniformly continuous on $[a, b]$, there exists $\delta > 0$ such that if $|t - t'| < \delta$, then

$$\|f(t) - f(t')\| < \frac{\varepsilon}{4(M - 1)}.$$

Now let $P_1 = \{a = t_0, t_1, \ldots, t_m = b\}$ be any partition of $[a, b]$ such that $\|P_1\| < \delta$. Let

$$V_1 = \sum_{k=1}^m \|f(t_k) - f(t_{k-1})\|.$$

Let $P = P_0 \cup P_1$ and let

$$V = \sum_P \|f(t_k) - f(t_{k-1})\|.$$

By the triangle inequality, we have $V \ge V_1$ and $V \ge V_0$. Suppose that P is formed by adding just one point $T_i \in P_0$ to P_1. Then $t_{j-1} < T_i < t_j$ for some

$j \le m$. Then

$$V = \sum_{k=1}^{j-1} \|f(t_k) - f(t_{k-1})\| + \|f(T_i) - f(t_{j-1})\| + \|f(t_j) - f(T_i)\|$$

$$+ \sum_{k=j+1}^{m} \|f(t_k) - f(t_{k-1})\|,$$

so that

$$V - V_1 = \|f(T_i) - f(t_{j-1})\| + \|f(t_j) - f(T_i)\| - \|f(t_j) - f(t_{j-1})\|$$

$$\le \|f(T_i) - f(t_{j-1})\| + \|f(t_j) - f(T_i)\|$$

$$\le 2\frac{\varepsilon}{4(M-1)} = \frac{\varepsilon}{2(M-1)}.$$

So adding one point to the partition P_1 increases V_1 by at most $\varepsilon/2(M-1)$. Since at most $M-1$ of the points in P_0 are different from the points in P_1, we conclude that

$$V - V_1 \le (M-1)\frac{\varepsilon}{2(M-1)} = \frac{\varepsilon}{2}.$$

Therefore,

$$V_1 \ge V - \frac{\varepsilon}{2} \ge V_0 - \frac{\varepsilon}{2} > T_f[a,b] - \varepsilon. \qquad \square$$

EXAMPLE 7.2.11. This example illustrates that the previous theorem does not hold in the case of discontinuous functions of bounded variation. Let $f: [0,1] \to \mathbf{R}^1$ such that

$$f(x) = \chi_{[1/2, 1]}(x).$$

Then f is not continuous and $T_f[0,1] = 1$. For any integer n, let P_n be a partition of $[0, 1]$ such that the component intervals of P_n have endpoints $k/2^n$ and $(k+1)/2^n$ for $0 \le k \le 2^n - 1$. Then $\|P_n\|$ is 2^{-n}, but

$$\sum_{P_n} |\Delta f_k| = 0.$$

Theorem 7.2.12 [J2, p. 567]. *Let $f, g: [a,b] \to \mathbf{R}^n$. Suppose each component function of f is absolutely continuous, g is of bounded variation, and $g' = 0$ a.e. Then*

$$T_{f+g}[a,b] = T_f[a,b] + T_g[a,b].$$

Proof. Let ϕ be defined on $[a,b]$ by

$$\phi(x) = T_{f+g}[a,x] - T_g[a,x].$$

Then, by Theorem 6.1.9,

$$T_{f+g}[a,b] - T_g[a,b] \leq T_f[a,b].$$

Since $g = (f + g) + (-f)$, we also have

$$T_g[a,b] - T_{f+g}[a,b] \leq T_{-f}[a,b] = T_f[a,b].$$

Therefore,

$$|T_{f+g}[a,b] - T_g[a,b]| \leq T_f[a,b].$$

In general, for $x < y$, we have

$$|\phi(y) - \phi(x)| = |T_{f+g}[a,y] - T_g[a,y] - T_{f+g}[a,x] + T_g[a,x]|$$
$$= |T_{f+g}[x,y] - T_g[x,y]| \leq T_f[x,y]$$
$$= T_f[a,y] - T_f[a,x].$$

By Theorem 7.1.14, $T_f[a,x]$ is absolutely continuous; therefore, ϕ is absolutely continuous. By Theorem 6.4.6, almost everywhere

$$\phi'(x) = \|f'(x) + g'(x)\| - \|g'(x)\| = \|f'(x)\| = T_f'[a,x].$$

Since $\phi(x)$ and $T_f[a,x]$ are absolutely continuous and have the same derivative a.e., we have

$$\phi(x) = T_f[a,x];$$

that is,

$$T_{f+g}[a,x] = T_g[a,x] + T_f[a,x].$$

Take $x = b$ to obtain the conclusion. □

EXAMPLE 7.2.13 [J2, p. 568]. In elementary calculus, we learn to calculate the length of curve by integrals. If $f : [a,b] \to \mathbf{R}^1$ has a continuous derivative, then the length of the curve from $(a, f(a))$ to $(b, f(b))$ is given by

$$\int_a^b \sqrt{1 + [f'(x)]^2}\, dt.$$

Let $f : [0,1] \to \mathbf{R}^1$ be a continuous function of bounded variation. We will determine the length of the curve $(t, f(t))$. By Theorem 7.1.46,

$$f(t) = F(t) + g(t),$$

where F is absolutely continuous and g is continuous and singular. Thus, the curve we are considering can be written as

$$(t, f(t)) = (t, F(t)) + (0, g(t)).$$

The length of the absolutely continuous part is

$$\int_0^1 \|(1, F'(t))\| \, dt = \int_0^1 \sqrt{1 + [F'(t)]^2} \, dt$$

$$= \int_0^1 \sqrt{1 + [f'(t)]^2} \, dt.$$

The length of the singular part is

$$T_{(0,g)}[0,1] = T_g[0,1].$$

Thus, the length of the graph $y = f(x)$ is given by

$$L = \int_0^1 \sqrt{1 + [f'(t)]^2} \, dt + T_g[0,1].$$

EXERCISES

1. Prove

$$f(x) = \begin{cases} x^2 \left| \sin \dfrac{1}{x} \right|, & x \neq 0 \\ 0, & x = 0, \end{cases}$$

   is AC (and non-negative) but $\sqrt{f}$ is not AC.

2. Let $f \in AC$. Prove $f(M)$ is measurable if M is measurable. [**Hint:** Use Theorem 7.1.38.]

3. Let

$$f(x) = x^{1/3},$$

$$h(x) = \begin{cases} x^3 \sin \dfrac{1}{x}, & x \neq 0 \\ 0, & x = 0. \end{cases}$$

   Show $f, h \in AC$ and $h \in C^1$, but $f \circ h$ is not AC and not even BV_{loc}.

4. If $f \in C^1$, show $f \in AC$.

5. If $f \in AC$, show $|f|^p \in AC$ for $p \geq 1$.

6. If $f(x) = |x|^p$, $p > 0$, then $f \in AC$.

7. Let

$$f(x) = \begin{cases} x^\alpha \sin x^{-\beta}, & x \neq 0 \\ 0, & x = 0, \end{cases}$$

   for $0 < \beta < \alpha$. Show $f \in AC$.

8. Let $f: \mathbf{R}^1 \to \mathbf{R}^1$ and let $\alpha > 0$. The function f satisfies a *Holder condition of order α* (also called a *Lipschitz condition of order α*) if there exists a constant c such

that

$$|f(x) - f(y)| \le c|x - y|^\alpha$$

for all $x, y \in \mathbf{R}^1$. Show if f satisfies a Holder condition of order $\alpha > 1$, then $f \in AC$.

9. Let f be AC on $\mathbf{R}^1 - \{0\}$. If $f \in BV_{loc}$ and f is continuous at 0, show $f \in AC$.

10. Let $\{f_n\} \in AC$ such that

$$\lim_{n, m \to \infty} T(f_n - f_m) = 0.$$

Show that exists $f \in AC$ such that

$$\lim_{n \to \infty} T(f - f_n) = 0.$$

11. Divide $[0, 1]$ by partition points 2^{-k} and let $I_k = (2^{-k}, 2^{-k+1})$. Divide I_k into 2^k equal subintervals I_{km}, $1 \le m \le 2^k$. On I_k let g be a non-negative, piecewise linear function which alternately takes the values 0 and 2^{-2k} at the left and right endpoints, respectively, of I_{km}. Show if $g(0) = 0$, then $g \in AC[0, 1]$. Also $f(x) = \sqrt{x}$ is absolutely continuous on $[0, 1]$. Show that $h = f \circ g$ is not AC. [**Hint**: Show that for any integer k, $T_h(I_k) = 1$.]

12. If f is continuous on $[a, b]$, if f' exists a.e. and is integrable on $[a, b]$, and if f maps null sets into null sets, then $f \in AC[a, b]$.

13. Let f be increasing on $[a, b]$ and let E be the subset where f' exists. Show that

$$\int_a^b f'(x)\, dx = m^*(f(E)) \le f(b) - f(a).$$

[**Hint**: Use Exercise 23 of Chapter 6 and Theorem 7.1.45.]

14. Let f be increasing on $[a, b]$, let A be a measurable subset, and let E be the subset of A where f' exists. Show that

$$\int_A f'(x)\, dx = m^*(f(E)) \le m^*(f(A)).$$

Show that if $f \in AC$, then the above becomes an equality. [**Hint**: First suppose $f \in AC$ and A is open, i.e., the union of a finite or countable collection of mutually disjoint open intervals. To extend the result to an arbitrary set A, consider sequences of open sets $\{A_n\}$ and $\{B_n\}$ such that $A \subset A_n$, $f(A) \subset B_n$, $\lim mA_n = mA$, and $\lim mB_n = m(f(A))$. Finally, in the general case, use Theorem 7.1.45.]

15. Let $f \in BV[a, b]$, let $V(x) = T_f[a, x]$, and let A be a measurable subset. Then

$$m^*(V(A)) \ge \int_A |f'(x)|\, dx,$$

with equality if $f \in AC[a, b]$. [**Hint**: Use Theorem 6.4.6 and Exercise 14.]

16. If f is increasing on $[a, b]$ and if M is any measurable subset for which $m(f(M)) = 0$, then $f'(x) = 0$ a.e. on M. [**Hint**: Use Exercises 13 and 14.]

17. Let f be continuous and of bounded variation on $[a, b]$ and let M be any subset for which $m(f(M)) = 0$. Show that $m(V(M)) = 0$, where V is defined as in Exercise 15. [**Hint**: Use Theorem 6.1.36.]

18. Let f be continuous and of bounded variation on $[a, b]$ and let M be a measurable subset for which $m(f(M)) = 0$. Then $f'(x) = 0$ a.e. on M. [**Hint**: Use Exercise 17 above and Exercise 25 of Chapter 6.]

19. A sequence $\{g_n\}$ in L^1_{loc} is said to converge to g in L^1_{loc} if, for each bounded interval $[a, b]$, $g_n \chi_{[a,b]} \to g \chi_{[a,b]}$ in L^1, i.e.,

$$\lim_{n \to \infty} \int_a^b |g(x) - g_n(x)| \, dx = 0.$$

Suppose $f, f_n \in AC$. Prove $T(f - f_n) \to 0$ if and only if $f'_n \to f'$ in L^1_{loc} and $f_n(0) \to f(0)$.

20. Verify that a summable function is asymptotically continuous at every Lebesgue point. If the function is bounded, the converse is also true.

21. Consider any interval $[a, b]$ and a perfect nowhere dense subset $F \subset [a, b]$ that contains both a and b; the measure of F may be zero or positive. Write $[a, b] \cap F^c$ as $\bigcup_{n=1}^\infty (a_n, b_n)$ where the intervals (a_n, b_n) are pairwise disjoint and are enumerated in an arbitrary order. Let $c_n = \frac{1}{2}(a_n + b_n)$ and let $\{t_n\}_{n=1}^\infty$ be a sequence of positive numbers with limit zero. Define a function g on $[a, b]$ as follows:

$g(x) = 0$ for all $x \in F$;
$g(c_n) = t_n$ $(n = 1, 2, \ldots)$;
g is linear in $[a_n, c_n]$ and in $[c_n, b_n]$ $(n = 1, 2, \ldots)$.

Prove that

(a) g is continuous;
(b) $T_g[a, b] = 2 \sum_{n=1}^\infty t_n$;
(c) g maps null sets into null sets.

Note if $\sum_{n=1}^\infty t_n = \infty$, then g is not BV, hence not AC.

22. Let E be a subset of $[0, 1]$ with measure zero. Construct an increasing and absolutely continuous function f on $[0, 1]$ such that $f'(x) = +\infty$ at every point $x \in E$.

CHAPTER 8

Cantor Sets and Singular Functions

In Chapter 7, we defined a singular function to be a nonconstant function whose derivative is zero a.e. The Cantor ternary function of Chapter 1 is an example of a continuous increasing singular function. In this chapter we consider singular functions in more detail. The study of such functions illustrates the beauty and subtlety of analysis in $\mathbf{R}^1$ as well as providing important applications to other fields, in particular, harmonic analysis. We begin the chapter with a recap of the results found in Chapter 1 on the Cantor ternary set and function, adding the result by Randolph that the set of distances between points of the Cantor set fills up the unit interval, some geometric results from Hille and Tamarkin, and a discussion of derivatives of the Cantor ternary function. In Section 8.2, we introduce Hausdorff measure and Hausdorff dimension and calculate the Hausdorff dimension of the Cantor ternary set. Sections 8.3, 8.4, and 8.5 present generalized Cantor sets and Cantor-like sets and their corresponding functions which share many, but not all, of the properties of the Cantor ternary set and function. A modulus of continuity result is given for Cantor-like sets in Section 8.5. The final section of this chapter gives details of the construction of two strictly increasing singular functions.

§8.1. The Cantor Ternary Set and Function

In Example 1.3.3, we began our discussion of the Cantor ternary function by first constructing the *Cantor ternary set*, denoted by C, through a process of removing intervals from the interval $[0, 1]$. By introducing ternary notation (base 3), we were able to characterize the numbers in the Cantor ternary set as those with ternary representations containing only the digits 0 and 2. Furthermore, points in C which are endpoints of the intervals removed have ternary representations where the digits after some point are all 0 or all 2. The other points in C are limit points of these endpoints and have ternary representations containing infinitely many 0's and 2's, except for the two extreme points 0 and 1. Thus, C^c, the complement (in $[0, 1]$) of C, consists of those numbers in the interval $[0, 1]$ for which no ternary representation has every digit either 0 or 2.

We now summarize some of the properties of the sets C and C^c.

(8.1.1)   Let $x \in C^c$, $x = .{}_3 a_1 a_2 \ldots a_k \ldots$. Then there exists a positive integer k such that if $a_k \neq 1$, then $a_i = 0$ or 2 for $i = 1, 2, \ldots, k - 1$; if $a_k = 1$, then not all $a_{k+1}, a_{k+2}, \ldots$ are equal to 0 or 2.

(8.1.2)   Let η_{pk} ($k = 1, \ldots, 2^p$) be the closed intervals remaining in $[0, 1]$ at the pth stage of the construction of C, i.e.,

$$\eta_{11} = [0, \tfrac{1}{3}], \qquad \eta_{12} = [\tfrac{2}{3}, 1], \quad \text{and so on.}$$

For fixed p, all the intervals η_{pk} have length 3^{-p} and the sum of their lengths is $(\tfrac{2}{3})^p$.

(8.1.3)   For any p, the intervals η_{pk} cover C; hence, C has measure zero.

(8.1.4)   Each point of C is a limit point of C; therefore, C is a perfect set.

(8.1.5)   Each subinterval of $(0, 1)$, no matter how small, contains points of C^c, i.e., C is nowhere dense.

(8.1.6 [G1])   The transformation

$$x' = 1 - x$$

leaves the sets C and C^c invariant.

Theorem 8.1.7 [R2]. *The set of distances between points of the Cantor set C fills the unit interval; that is, given $c \in [0, 1]$, there exist $x, y \in C$ such that $y - x = c$.*

Proof.   Let $c \in [0, 1]$ be written in its ternary representation, i.e., $c = .{}_3 \gamma_1 \gamma_2 \gamma_3 \ldots$ where each $\gamma_i = 0$, 1, or 2. We will show that there exists $x = .{}_3 \alpha_1 \alpha_2 \alpha_3 \ldots$ and $y = .{}_3 \beta_1 \beta_2 \beta_3 \ldots$, where each α_i and β_i is either 0 or 2 unless α_i (β_i) is the last significant digit of x (y) in which case it may be 1, such that $y = x + c$.

Let δ denote a succession of 0's and 2's or contains no elements. For example,

$$.{}_3 021220110200 \ldots$$

is written as

$$.{}_3 \delta_1 1 \delta_2 1 \delta_3 1 \delta_4$$

where $\delta_1 = 02$, $\delta_2 = 220$, $\delta_3 = \varnothing$, and $\delta_4 = 0200\ldots$. For any known δ_i, let δ_i' be obtained from δ_i by interchanging the digits 0 and 2. Thus, in the above example, $\delta_2' = 002$. Let $(0)_i$ be the succession of 0's obtained from δ_i by keeping each 0 intact and replacing each 2 by 0.

Case 1: Let $c = {}_{.3}\gamma_1\gamma_2\gamma_3\ldots$ contain the digit 1 an even (hence finite) number of times or not at all. Then

$$c = {}_{.3}\delta_1 1\delta_2 1\ldots\delta_{2n-1} 1\delta_{2n} 1\delta_{2n+1}.$$

Let

$$x = {}_{.3}(0)_1 0\delta_2' 2\ldots(0)_{2n-1} 0\delta_{2n}' 2(0)_{2n+1},$$

$$y = {}_{.3}\delta_1 2(0)_2 0\ldots\delta_{2n-1} 2(0)_{2n} 0\delta_{2n+1}.$$

Then $x, y \in [0, 1]$. Note that $\delta_{2n+1} + (0)_{2n+1} = \delta_{2n+1}$. Adding the last 1 of c to the last 2 of x, we get 0 with 1 to carry. The sum of the last digits of δ_{2n} and δ_{2n}' is 2. With the 1 on the carryover, we get 0 as long as the digits of δ_{2n} last. We then have 1 to carry over to the digit 1 before δ_{2n+1} which, with the 0 before δ_{2n}' of x gives 2 with none to carry. Proceed with δ_{2n-1} exactly as we started with δ_{2n+1}. Thus $y = x + c$.

Case 2: The digit 1 appears an odd finite number of times among the γ_i of c, i.e.,

$$c = {}_{.3}\delta_1 1\delta_2 1\ldots\delta_{2n-1} 1\delta_{2n} 1\delta_{2n+1} 1\delta_{2n+2}.$$

Let

$$x = {}_{.3}(0)_1 0\delta_2' 2\ldots(0)_{2n-1} 0\delta_{2n}' 2(0)_{2n+1} 1(0)_{2n+2},$$

$$y = {}_{.3}\delta_1 2(0)_2 0\ldots\delta_{2n-1} 2(0)_{2n} 0\delta_{2n+1} 2\delta_{2n+2}.$$

(The digit 1 is permissible in x since it is the last significant digit.) The only difference in x and y from Case 1 is in the two symbols.

Case 3: The digit 1 occurs an infinite number of times in c, i.e.,

$$c = {}_{.3}\delta_1 1\delta_2 1\delta_3 1\delta_4 1\ldots.$$

Let

$$x = {}_{.3}(0)_1 0\delta_2' 2(0)_3 0\delta_4' 2\ldots,$$

$$y = {}_{.3}\delta_1 2(0)_2 0\delta_3 2(0)_4 0\ldots,$$

$$c_n = {}_{.3}\delta_1 1\delta_2 1\ldots\delta_{2n-1}\delta_{2n}.$$

Then

$$\lim_{n\to\infty} c_n = c.$$

Similarly, define x_n and y_n. Note that by case 1, $y_n = x_n + c_n$. Thus, $y = x + c$.

$\square$

Remark. Given $c \in (0, 1)$, the choice of x and y in the above theorem is not necessarily unique. For example, if $c = .{_3}021201$, then we have

$$.{_3}021201 + .{_3}001000 = .{_3}022201,$$

$$.{_3}021201 + .{_3}201000 = .{_3}222201$$

$$.{_3}021201 + .{_3}000022 = .{_3}022000.$$

But $.{_3}021201 + .{_3}202001 = 1.{_3}000202$ and we get x and y whose fractional parts may be written using only 0's or 2's, but $y > 1$ and hence is not in the Cantor set.

We now restate the definition of the *Cantor function* ω from Example 1.3.3 and list some of its properties.

Definition 8.1.8. For $x = .{_3}a_1 a_2 \ldots \in C$ (where a_i is either 0 or 2), let $b_i = a_i/2$, so that b_i assumes only the values 0 or 1. Then for $x \in C$, we define $\omega(x) = .{_2}b_1 b_2 \ldots$.

(8.1.9) ω has the same value at the endpoints of each E_{pk} $(k = 1, \ldots, 2^{p-1})$.

(8.1.10)   If we define ω to be constant in the interval E_{pk}, then $\omega(x)$ is now defined for all $x \in [0, 1]$.

(8.1.11) ω increases from 0 to 1 as x increases from 0 to 1.

Theorem 8.1.12 [G1]. $\omega(x) + \omega(1 - x) = 1$.

Proof. By (8.1.6), if $x \in C$, then $x' = 1 - x \in C$. Let

$$x = \frac{a_1}{3} + \frac{a_2}{3^2} + \cdots,$$

$$x' = \frac{a_1'}{3} + \frac{a_2'}{3^2} + \cdots.$$

Clearly, $a_i' = 2 - a_i$, which implies that

$$h_i' = \frac{a_i'}{2} = \frac{2 - a_i}{2} = 1 - b_i.$$

Thus,

$$\omega(x) = \frac{b_1}{2} + \frac{b_2}{2^2} + \cdots,$$

$$\omega(x') = \frac{1 - b_1}{2} + \frac{1 - b_2}{2^2} + \cdots.$$

Therefore, $\omega(x) + \omega(x') = 1$ for $x \in C$. If $x \in C^c$, $x = .{_3}a_1 a_2 \ldots$, then there exists a positive integer k such that for $i \le k - 1$, we have $a_i' = 2 - a_i$, $b_i' = 1 - b_i$, and $b_k + b_k' = 1$. The sequences $\{b_n\}$ and $\{b_n'\}$ terminate at this stage. Thus, $\omega(x) + \omega(x') = 1$. $\square$

(8.1.13) ω is continuous on $[0, 1]$.

(8.1.14) ω is *not* absolutely continuous on $[0, 1]$.

Proof. In Chapter 1, we proved that $\omega' = 0$ a.e. and $\omega(1) - \omega(0) = 1$. Therefore,

$$\int_0^1 \omega'(x)\, dx = 0 \ne 1 = \omega(1) - \omega(0).$$

Thus, ω is not absolutely continuous by Theorem 7.1.15. $\square$

Definition 8.1.15 [HT]. We define the λ-*variation* of a function f on an interval $[a, b]$ to be the upper limit of the sums

$$\sum_{k=1}^m |f(\beta_k) - f(\alpha_k)|$$

where $\{(\alpha_k, \beta_k)\}$ is any finite set of nonoverlapping intervals of $[a, b]$ such that

$$\sum_{k=1}^m (\beta_k - \alpha_k) \le \lambda.$$

Theorem 8.1.16 [HT]. *For every positive λ, the λ-variation of ω is 1.*

Proof. Let $(\alpha_k, \beta_k) = \eta_{pk}$. Then

$$\sum_{k=1}^m |\omega(\beta_k) - \omega(\alpha_k)| = 1.$$

Since

$$\sum_{k=1}^m (\beta_k - \alpha_k) = \sum_{k=1}^m m(\eta_{pk})$$

can be made as small as we choose by taking p large, the result follows. This result also provides another proof that ω is not absolutely continuous. $\square$

(8.1.17 [HT]) The function

$$y = \psi(x) = \frac{x + \omega(x)}{2}$$

gives a continuous one-to-one correspondence between the segment $[0, 1]$ on the x axis and the segment $[0, 1]$ on the y axis.

(8.1.18 [HT]) The function given in (8.1.17) transforms each interval E_{pk} on the x axis into an interval of length $mE_{pk}/2$ on the y axis. Thus, C^c with measure 1 is transformed into a set of measure $\frac{1}{2}$. It follows that C, of measure 0, is transformed into a set of measure $\frac{1}{2}$. Since every set of positive measure contains a nonmeasurable subset, the function ψ provides an example of a continuous one-to-one transformation in which a measurable set (even of measure zero) is transformed into a nonmeasurable set.

Theorem 8.1.19 [HT]. *The area under the curve $y = \omega(x)$ is $\frac{1}{2}$.*

Proof.

$$\int_0^1 \omega(x)\,dx = \int_C \omega(x)\,dx + \int_{C^c} \omega(x)\,dx$$

$$= \int_{C^c} \omega(x)\,dx = \sum_{p=1}^{\infty} \sum_{k=1}^{2^{p-1}} \omega(x)m(E_{pk})$$

$$= \sum_{p=1}^{\infty} 3^{-p} \sum_{k=1}^{2^{p-1}} (2k-1)2^{-p} = \sum_{p=1}^{\infty} 6^{-p} \sum_{k=1}^{2^{p-1}} (2k-1)$$

$$= \sum_{p=1}^{\infty} 6^{-p}2^{2p-2} = \frac{1}{4} \sum_{p=1}^{\infty} \left(\frac{2}{3}\right)^p = \frac{1}{6}\left(\frac{1}{1-\frac{2}{3}}\right) = \frac{1}{2}. \qquad \square$$

Theorem 8.1.20 [HT]. *The length of the arc of the curve $y = \omega(x)$ between $(0,0)$ and $(1,1)$ is 2.*

Proof. Since ω is monotone, it is of bounded variation and hence the curve under consideration has finite length, which is the limit of the perimeters of inscribed polygons. We note that the perimeter of any inscribed polygon which does not cross itself cannot exceed the sum of all the lengths of the horizontal and vertical projections of its sides, which will equal 2, so long as the polygon starts at $(0,0)$ and ends at $(1,1)$. Thus, the perimeter of any inscribed polygon not crossing itself is less than or equal to 2. Now consider the polygon formed by the broken line whose vertices are at $(0,0)$, $(1,1)$, and at the endpoints of E_{pk} [$p = 1, 2, \ldots, n$ (fixed), $k = 1, 2, \ldots, 2^{p-1}$]. The sum of the horizontal sides of this polygon is

$$\sum_{p,k} mE_{ph} = \sum_{p=1}^{n} 2^{p-1}3^{-p} = 1 - \left(\frac{2}{3}\right)^n.$$

The inclined sides are all equal and their common length is

$$(2^{-2n} + 3^{-2n})^{1/2} = 2^{-n}[1 + \left(\frac{2}{3}\right)^{2n}]^{1/2}.$$

The number of such sides is 2^n. Thus, the length of this polygon is

$$1 - \left(\frac{2}{3}\right)^n + [1 + \left(\frac{2}{3}\right)^{2n}]^{1/2} \to 2 \quad \text{as } n \to \infty.$$

It should be noted here that since ω is not absolutely continuous, the length of the curve cannot be computed by the formula

$$\int_0^1 [1 + \omega'(x)^2]^{1/2}\, dx = 1 \neq 2.$$

(Compare to Example 7.2.13.) □

Theorem 8.1.21 [HT]. *For $x, x + h \in [0, 1]$,*

$$|\omega(x + h) - \omega(x)| \leq \Lambda |h|^{\alpha},$$

where $1 \leq \max \Lambda \leq 2$ and $\alpha = (\log 2)/(\log 3)$.

Proof. By choosing $[x, x + h] = \eta_{pk}$, we have $\omega(x + h) - \omega(x) = 2^{-p}$ and $h = 3^{-p}$. Thus,

$$|\omega(x + h) - \omega(x)| = 2^{-p} \leq \Lambda_1 |h|^{\alpha_1}$$

is not possible for any $\alpha_1 < (\log 2)/(\log 3)$ and any λ_1 such that $\max \lambda_1 < 1$. In this sense these are the best possible values for Λ and α. To prove Theorem 8.1.21 we can restrict our attention to the case $x, x + h \in C$; otherwise, we can replace x or $x + h$ by the appropriate endpoints for the corresponding interval E_{pk}. This will not change the value of $\omega(x + h) - \omega(x)$ but will reduce h. Let

$$x = {}_{.3}a_1 a_2 \ldots,$$

$$x + h = {}_{.3}a_1' a_2' \ldots.$$

Assume $h > 0$. Then there exists a positive integer n such that

$$a_1 = a_1', \ldots, a_{n-1} = a_{n-1}', \quad a_n' > a_n.$$

This implies that $a_n' = 2$ and $a_n = 0$. Then

$$\omega(x + h) - \omega(x) = {}_{.2}b_1 b_2 \ldots b_{n-1} 1 \ldots - {}_{.2}b_1 b_2 \ldots b_{n-1} 0 \ldots \leq 2^{-n+1},$$

whereas

$$h = {}_{.3}a_1 a_2 \ldots a_{n-1} 2 \ldots - {}_{.3}a_1 a_2 \ldots a_{n-1} 0 \ldots \geq 3^{-n}.$$

Therefore,

$$[\omega(x + h) - \omega(x)]h^{-\alpha} \leq 2^{-n+1} 3^{\alpha n} = 2$$

if $\alpha = (\log 2)/(\log 3)$. □

Now let us consider the derivatives of the Cantor ternary function ω. At the left endpoint of one of the intervals in the complement of C, f_+' exists and is zero. Similarly, $f_-'(x) = 0$ at a right-hand endpoint x of such an interval. Now we consider the derivatives at the other sides of complementary intervals. For the sake of definiteness, let a right-hand endpoint be

$$x = {}_{.3}a_1 a_2 \ldots a_n 2000 \ldots$$

If $3^{-m} < h < 3^{-m-1}$, $m > n$, $\omega(x + h)$ differs from $\omega(x)$ by some value between 2^{-m} and 2^{-m-1}. So

(8.1.22)
$$\frac{\omega(x + h) - \omega(x)}{h}$$

must be between $2^{-m-1}/3^{-m}$ and $2^{-m}/3^{-m-1}$. Thus, as $h \to 0$, $m \to \infty$ and (8.1.22) becomes $+\infty$. Therefore, at a right-hand endpoint x, $f'_+(x) = +\infty$ and $f'_-(x) = 0$. Similarly, $f'_-(x) = +\infty$ and $f'_+(x) = 0$ at a left-hand endpoint. Proceeding similarly at a limit point that is not the endpoint of such an interval, we get $D^+ f(x) = +\infty$ while $D_+ f(x)$ can have any value between 0 and $+\infty$.

EXAMPLE 8.1.23. We now give an example of two functions which have the same derivative (possibly infinite) but which do not differ from each other by a constant (cf. Section 3.3). Let g be defined as the Cantor ternary function, with the exception that in each interval on which ω is a constant, the graph of g is two symmetric semicircles, one above the graph of ω and the other below. Then $g'_- = +\infty$ at the right-hand endpoints of such intervals and $g'_+ = +\infty$ at the left-hand endpoints. Thus, g' exists throughout $(0, 1)$ and $g' = +\infty$ at all points of the Cantor set, whereas $g' = -\infty$ at the midpoints of the complementary intervals.

Let $h = \omega + g$. Then $h' = g'$ at points of the Cantor set. In the interior of the complementary intervals $h' = g'$ (because $\omega' = 0$ there). Thus, g and h have the same derivative (possibly infinite) but $h - g = \omega$, which is not a constant function.

EXAMPLE 8.1.24. Since the Cantor ternary set C contains points whose ternary representations are of the form

$$.3 a_1 a_2 a_3 \ldots$$

where each a_i is either 0 or 2, any $x \in C$ can be represented in the form

$$x = \sum_{n=1}^{\infty} \varepsilon_n \frac{2}{3^n},$$

where each ε_n is either 0 or 1. Now we define a function $\phi: C \to [-\frac{1}{2}, \frac{1}{2}]$ such that

$$\phi\left(\sum_{n=1}^{\infty} \varepsilon_n \frac{2}{3^n}\right) = -\frac{1}{2} + \sum_{n=1}^{\infty} \varepsilon_n \frac{2}{3^n}.$$

Note that $\phi(C) = [-\frac{1}{2}, \frac{1}{2}]$. If $\gamma: (-\frac{1}{2}, \frac{1}{2}) \to \mathbf{R}^1$ by $\gamma(x) = \tan \pi x$, then γ is continuous, the image of γ is $\mathbf{R}^1$, $h_0 = \gamma \circ \phi$ is defined on $D_0 = C - \{0, 1\}$, and the image of h_0 is $\mathbf{R}^1$. Recall that $[0, 1] - C$ is the union of countably many pairwise disjoint open intervals; we can repeat a similar construction for each of these intervals to get a sequence of sets $\{D_n\}$, each of which is the analogue of D_0 on $[0, 1]$. Let

$$D = \bigcup_{n=0}^{\infty} D_n.$$

Note that the D_i's are disjoint and $mD_n = 0$, $n \geq 0$; thus, $mD = 0$. For each D_n, define h_n as above. Let

$$H(x) = \begin{cases} 0, & x \notin D \\ h_n(x), & x \in D_n, n \geq 0. \end{cases}$$

Then $H = 0$ a.e. Since inside every nonempty open interval $(a, b) \subset [0, 1]$, there exists an interval in which some D_n lies and $h_n(D_n) = \mathbf{R}^1$, it follows that $H([a, b]) = \mathbf{R}^1$. Thus, $H = 0$ a.e., but H maps every nondegenerate interval $(a, b) \subset [0, 1]$ *onto* $\mathbf{R}^1$.

§8.2. Hausdorff Measure

A function f, defined on $[a, b]$, is said to satisfy a *uniform Lipschitz condition of order* $\alpha > 0$ (also called a *Holder condition of order* α) if there exists a constant $M > 0$ such that

$$|f(x) - f(y)| \leq M|x - y|^{\alpha}$$

for all $x, y \in [a, b]$. We see from Theorem 8.1.21 that the Cantor ternary function ω satisfies such a condition with $\alpha = (\log 2)/(\log 3)$. Interestingly, this number α also has a special significance with regard to the Cantor ternary set C. To explore that significance, we need to introduce the concept of *Hausdorff measure* and *Hausdorff dimension*. Using Hausdorff measure, we can determine in some sense how "big" a set with zero Lebesgue measure is. In this section, we will present the main properties of Hausdorff measure on the real line; our chief source for these results is [D1].

Recall from Chapter 0 that the outer Lebesgue measure of a set $E \subset \mathbf{R}^1$ is given by

$$m^*(E) = \inf_{E \subset \bigcup_n I_n} \sum_n |I_n|,$$

where each $\{I_n\}$ is a Lebesgue covering of E, i.e., a countable collection of open intervals such that

$$E \subset \bigcup_n I_n,$$

and $|\cdot|$ denotes length.

Definition 8.2.1. Let $s \geq 0$. For $\delta > 0$, define the "approximating measure" H_δ^s of the set $E \subset \mathbf{R}^1$ by

$$H_\delta^s(E) = \inf_{E \subset \bigcup_n I_n, |I_n| \leq \delta} \sum_n |I_n|^s.$$

Definition 8.2.2. The *s-dimensional Hausdorff outer measure* of E is given by

$$H^s(E) = \sup_{\delta \geq 0} H^s_\delta(E) = \lim_{\delta \to 0+} H^s_\delta(E).$$

We justify the last equality by observing that the effect of reducing δ is to reduce the class of coverings over which the infimum is taken in Definition 8.2.1; thus, $H^s_\delta(E)$ can only increase as δ decreases.

Theorem 8.2.3.

(a) $H^s(E) \geq 0$,
(b) $H^s(\varnothing) = 0$,
(c) $H^s(A) \leq H^s(B)$ if $A \subset B$,
(d) $H^s(\{x\}) = 0$ for any $x \in \mathbf{R}^1$.

Proof. These results follow easily from the definition.   ☐

Theorem 8.2.4. *Hausdorff outer measure is invariant under translation, i.e.,* $H^s(E) = H^s(E + x)$, *where $E + x$ is the set $\{y + x : y \in E\}$.*

Proof. Each covering of E by intervals $\{I_n\}$ of length at most δ corresponds to a cover $\{I_n + x\}$ of $E + x$ and

$$\sum_n |I_n|^s = \sum_n |I^n + x|^s.$$

Taking the infimum over such covers gives

$$H^s_\delta(E) = H^s_\delta(E + x).$$

Letting $\delta \to 0$ yields the desired result.   ☐

Theorem 8.2.5. $H^s(kE) = k^s H^s(E)$ *for any positive k.*

Proof. Each covering of E by intervals $\{I_n\}$ of length at most δ corresponds under the map $x \to kx$ to a cover of kE by intervals $\{kI_n\}$ of length at most $k\delta$ and

$$\sum |kI_n|^s = k^s \sum |I_n|^s.$$

Taking the infimum over all such covers $\{I_n\}$ gives $H^s_{k\delta}(kE) = k^s H^s_\delta(E)$. The desired results follows by letting $\delta \to 0$.   ☐

Theorem 8.2.6. $H^s(E)$ *is the same whether we require the coverings $\{I_n\}$ to be open, closed, or half-open. In fact, we obtain the same value for $H^s(E)$ if the covering are arbitrary sets $\{U_n\}$, replacing $|I_n|^s$ and $|I_n| < \delta$ by $[\mathrm{diam}(U_n)]^s$ and $\mathrm{diam}(U_n) < \delta$, respectively, in Definition 8.2.1.*

Proof. We will prove that closed intervals give the same outer measure, the other cases being similar. Let H_δ^{sc} and H^{sc} represent the approximating measure and Hausdorff measure, respectively, obtained from coverings by closed intervals. Let $\bar{I}$ represent the closure of the open interval I and let $\{J_n\}$ be a collection of closed intervals. Then

$$H_\delta^s(E) = \inf_{|I_n| \le \delta, E \subset \bigcup_n I_n} \sum_n |I_n|^s$$

$$= \inf \sum_n |\bar{I}_n|^s$$

$$\ge \inf_{|J_n| \le \delta, E \subset \bigcup_n J_n} \sum_n |J_n|^p$$

$$\ge H_\delta^{sc}(E).$$

So $H^s(E) \ge H^{sc}(E)$. To prove the opposite inequality, we observe that every closed interval J of length $\varepsilon > 0$ is contained in an open interval J' of length $\varepsilon(1 + \delta)$. Then

$$H_\delta^{sc}(E) = \inf_{|J_n| \le \delta, E \subset \bigcup_n J_n} \sum_n |J_n|^s$$

$$= (1 + \delta)^{-s} \inf \sum_n |J_n'|^s$$

$$\ge (1 + \delta)^{-s} \inf_{|I_n| \le \delta + \delta^2, E \subset \bigcup_n I_n} \sum_n |I_n|^s$$

$$= (1 + \delta)^{-s} H_{\delta+\delta^2}^s(E).$$

Let $\delta \to 0$ to get $H^{sc}(E) \ge H^s(E)$. □

Theorem 8.2.7. Let $H^s(E) < \infty$ and let $q > s$. Then $H^q(E) = 0$.

Proof. Let $\delta > 0$ and let $\{I_n\}$ be a covering of E by intervals such that $|I_k| \le \delta$ for each n. Then

$$\frac{|I_n|^q}{|I_n|^s} = |I_n|^{q-s} \le \delta^{q-s}.$$

Therefore,

$$H_\delta^q(E) \le \sum_n |I_n|^q \le \delta^{q-s} \sum_n |I_n|^s.$$

Taking the infimum of the last term over all such coverings, we get

$$H_\delta^q(E) \le \delta^{q-s} H_\delta^s(E) \le \delta^{q-s} H^s(E).$$

Letting $\delta \to 0$ yields the desired result. □

Corollary 8.2.8. If $0 < H^s(E) < \infty$, then $H^q(E) = \infty$ for $q < s$.

Theorem 8.2.9. For any sequence of sets $\{E_n\}$, we have

$$H^s\left(\bigcup_{n=1}^\infty E_n\right) \le \sum_{n=1}^\infty H^s(E_n).$$

Proof. For every $\varepsilon > 0$, there exists a family of intervals $\{I_{n,j}\}$ with $|I_{n,j}| \leq \delta$ such that

$$E_n \subset \bigcup_{j=1}^{\infty} I_{n,j}$$

for each n and

$$H_\delta^s(E_n) \geq \sum_{n=1}^{\infty} |I_{n,j}|^s - \frac{\varepsilon}{2^n}.$$

Then

$$\bigcup_{n=1}^{\infty} E_n \subset \bigcup_{n=1}^{\infty} \bigcup_{j=1}^{\infty} I_{n,j},$$

so

$$H_\delta^s\left(\bigcup_{n=1}^{\infty} E_n\right) \leq \sum_{n,j} |I_{n,j}|^s$$

$$\leq \sum_n H_\delta^s(E_n) + \varepsilon$$

$$\leq \sum_n H^s(E_n) + \varepsilon,$$

for all $\delta > 0$. Therefore,

$$H^s\left(\bigcup_{n=1}^{\infty} E_n\right) \leq \sum_n H^s(E_n) + \varepsilon.$$

Since ε is arbitrary, the result follows.   □

Remark. From the definition of Hausdorff measure, it is clear that for any set $A \subset \mathbf{R}^1$, $H^1(A) = m^*(A)$.

Theorem 8.2.10. *Let $I \subset \mathbf{R}^1$ be an interval of positive or infinite length. Then $H^s(I) = \infty$ for $0 < s < 1$, $H^1(I) = |I|$, and $H^s(I) = 0$ for $s > 1$.*

Proof. Suppose that I is a finite interval. Then the result follows from Theorem 8.2.7, Corollary 8.2.8, and Theorem 8.2.9. So suppose that I is an infinite interval and $s > 1$. Then

$$I \subset \bigcup_{k=1}^{\infty} I_k,$$

where each I_k is a finite interval. Thus, we have

$$H^s(I) \leq \sum_k H^s(I_k) = 0.$$

So the result is true for $s > 1$; by Corollary 8.2.8, we have $H^s(I) = \infty$ for $0 < s \leq 1$.   □

Corollary 8.2.11. $H^s(\mathbf{R}^1) = \infty$ *for* $0 < s \leq 1$; $H^s(\mathbf{R}^1) = 0$ *for* $s > 1$.

Corollary 8.2.12. *If* $s < 1$, $H^s(G) = \infty$ *for every nonempty open set* $G \subset \mathbf{R}^1$.

Corollary 8.2.13. *For a fixed set* $A \subset \mathbf{R}^1$, *consider* $H^s(A)$ *as a function of* s *($s > 0$). Either* $H^s(A) = 0$ *for all* $s > 0$, *or for some* s_0 *($0 < s_0 \leq 1$) we have* $H^s(A) = \infty$ *for* $0 < s < s_0$ *and* $H^s(A) = 0$ *for* $s > s_0$.

Theorem 8.2.14. *The H^s-measurable sets form a σ-algebra.*

Proof. The proof is the same as for Lebesgue measure. □

Theorem 8.2.15. *H^s is countably addditive on the σ-algebra of H^s-measurable sets.*

Proof. Again the proof is the same as for Lebesgue measure. □

Theorem 8.2.16. *H^s is a* metric outer measure; *that is, if A and B are nonempty disjoint sets in $\mathbf{R}^1$ with* $\text{dist}(A, B) > 0$, *then* $H^s(A \cup B) = H^s(A) + H^s(B)$.

Proof. It is sufficient to show that the same identity holds for H^s_δ for all small δ. But this result is immediate for H^s_δ if $\delta < \text{dist}(A, B)$ because any covering of $A \cup B$ by intervals of length at most δ decomposes into two coverings of A and B. □

Theorem 8.2.17. *Let I be the interval $(-\infty, a]$. Then I is H^s measurable.*

Proof. We need to show that for any set A,

(8.2.18) $$H^s(A) \geq H^s(A \cap I) + H^s(A - I).$$

We may suppose that $H^s(A)$ is finite; otherwise (8.2.18) follows immediately. Let

$$A_n = A \cap \left[a + \frac{1}{n}, \infty \right).$$

Then $A_n \subset A_{n+1}$ and

$$\bigcup_{n=1}^{\infty} A_n = A - I.$$

Also,

$$\lim_{n \to \infty} H^s(A_n)$$

exists and is finite. By Theorem 8.2.16, we have

$$H^s(A) \geq H^s(A \cap I) + H^s(A_n).$$

So if we show that

$$\lim_{n \to \infty} H^s(A_n) = H^s(A - I),$$

the result will follow. Let $D_n = A_{n+1} - A_n$. Then

$$A - I = A_{2n} \cup \left(\bigcup_{k=2n}^{\infty} D_k \right) = A_{2n} \cup \left(\bigcup_{k=n}^{\infty} D_{2k} \right) \cup \left(\bigcup_{k=n}^{\infty} D_{2k+1} \right).$$

So

(8.2.19) $H^s(A - I) \leq H^s(A_{2n}) + \sum_{k=n}^{\infty} H^s(D_{2k}) + \sum_{k=n}^{\infty} H^s(D_{2k+1}),$

and all these outer measures are finite. Suppose that both series in (8.2.19) converge. Then letting $n \to \infty$, we get $H^s(A - I) \leq \lim H^s(A_{2n}) = \lim H^s(A_n)$. But since $A_n \subset A - I$, we must have equality and (8.2.18) follows. Suppose that the first series in (8.2.19) diverges. We have

$$\bigcup_{k=1}^{n-1} D_{2k} \subset A_{2n},$$

and $\text{dist}(D_{2k}, D_{2k+1}) > 0$. Therefore, by Theorem 8.2.16, we have

$$H^s(A_{2n}) \geq \sum_{k=1}^{n-1} H^s(D_{2k}) \to \infty,$$

contradicting the finiteness of $H^s(A)$. Similarly, the second series in (8.2.19) must converge and the desired result follows. $\square$

Corollary 8.2.20. *All Borel sets are H^s-measurable.*

Definition 8.2.21. The *Hausdorff dimension* of a set E is $\inf\{s: H^s(E) = 0\}$.

By Corollary 8.2.13, Hausdorff dimension is a well-defined number in the interval $[0, 1]$ for any set $A \subset \mathbf{R}^1$. Also, except for the trivial case of sets E for which $H^s(E) = 0$ for all s, we have the Hausdorff dimension of a set E is $\sup\{s: H^s(E) = \infty\}$.

Theorem 8.2.22. *The Hausdorff dimension of the Cantor ternary set C is $(\log 2)/(\log 3)$.*

Proof. Refer to the construction of the Cantor ternary set in Example 1.3.3. At the first step of the construction, we removed the open interval $E_{11} = (\frac{1}{3}, \frac{2}{3})$, leaving the two closed intervals $[0, \frac{1}{3}]$ and $[\frac{2}{3}, 1]$. Each of these closed intervals are translates of multiples of $[0, 1]$ and contain subsets, call them $C^{(1)}$ and $C^{(2)}$, of C. We could have constructed $C^{(1)}$ in the same manner as we constructed C, beginning with the interval $[0, \frac{1}{3}]$ instead of $[0, 1]$. Thus, we have that $C^{(1)}$ (and $C^{(2)}$) is a translate of a multiple (actually $\frac{1}{3}$) of C. Then by

Theorems 8.2.4, 8.2.5, and 8.2.16,

$$H^s(C) = H^s(C^{(1)} \cup C^{(2)}) = H^s(C^{(1)}) + H^s(C^{(2)}) = 2(\tfrac{1}{3})^s H^s(C).$$

So either $H^s(C) = 0$ or ∞, or $2(\tfrac{1}{3})^s = 1$. In the latter case, $s = (\log 2)/(\log 3) = s_0$. If we can show that $0 < H^{s_0}(C) < \infty$, it will follow that C has Hausdorff dimension s_0.

First, we will show that $H^{s_0}(C) < \infty$. Let $\{I_i\}$ be a covering of C by open intervals of length at most δ. As above, we construct sets $C^{(j)}$ ($j = 1, 2$) and use the same kind of similarity transformation on the intervals $\{I_{i,j}\}$ to construct coverings $\{J_{i,j}\}$ for $C^{(j)}$ for $j = 1, 2$. Since $2(\tfrac{1}{3})^{s_0} = 1$,

(8.2.23) $|I_i|^{s_0} = 2(\tfrac{1}{3})^{s_0}|I_i|^{s_0} = |J_{i,1}|^{s_0} + |J_{i,2}|^{s_0}.$

Taking the infimum over all such coverings $\{I_i\}$ of C, we get

$$H^{s_0}_\delta(C) \geq H^{s_0}_{\delta/3}(C)$$

since $|J_{i,j}| \leq \tfrac{1}{3}\delta$. Since H^s_δ does not decrease as δ decreases and $0 < \tfrac{1}{3} < 1$, it follows that $H^{s_0}_\delta$ is independent of δ. Thus, we may take as a cover an open interval of length just greater than 1 and containing C to get $H^{s_0}(C) \leq 1$.

Now we show that $H^{s_0}(C) > 0$. The distance between the sets $C^{(1)}$ and $C^{(2)}$ is at least $1 - 2(\tfrac{1}{3})$. Let $\delta \leq 1 - 2(\tfrac{1}{3})$. Then, as in Theorem 8.2.16, any cover $\{I_i\}$ of C by intervals of length at most δ may be decomposed into two covers $\{I_{i,j}\}$, $j = 1, 2$, of the sets $C^{(1)}$ and $C^{(2)}$ and

(8.2.24) $\sum_i |I_i|^{s_0} = \sum_i |I_{i,1}|^{s_0} + \sum_i |I_{i,2}|^{s_0}.$

Suppose that the first sum on the right of (8.2.24) is the smaller. Since $C^{(2)}$ is a translate of $C^{(1)}$, the same translation applied to the intervals $\{I_{i,1}\}$ gives a cover $\{I'_{i,1}\}$ of $C^{(2)}$. Then in a "reverse"-type transformation, we map the intervals $\{I_{i,1}\}$ onto intervals $\{I'_i\}$ covering C with $\tfrac{1}{3}|I'_i| = |I_{i,1}|$ for each i. As in (8.2.23), we have

(8.2.25) $\sum_i |I'_i|^{s_0} = \sum_i |I_{i,1}|^{s_0} + \sum_i |I'_{i,1}|^{s_0} \leq \sum_i |I_i|^{s_0}.$

So if any one of the intervals I'_i is of length $\geq 1 - 2(\tfrac{1}{3})$, we have

$$\sum_i |I_i|^{s_0} \geq [1 - 2(\tfrac{1}{3})]^{s_0}.$$

Since C is compact, we may suppose that the coverings considered are finite, so that $\min|I_k| > 0$. Since the intervals $\{I'_i\}$ are multiples [by $(\tfrac{1}{3})^{-1}$] of a subset of the intervals $\{I_k\}$, we have

(8.2.26) $\min|I'_i| \geq (\tfrac{1}{3})^{-1} \min|I_k|.$

If each interval I'_i is of length less than $1 - 2(\tfrac{1}{3})$, we apply the same process to the cover $\{I'_i\}$ which was applied to $\{I_i\}$. After a finite number of steps, we must obtain a cover $\{I''_i\}$ with $\max|I''_i| \geq 1 - 2(\tfrac{1}{3})$ and

$$\sum_i |I''_i|^{s_0} \leq \sum_i |I_i|^{s_0}$$

as in (8.2.24). In any case we have

$$\sum_i |I_i|^{s_0} \geq [1 - 2(\tfrac{1}{3})]^{s_0}.$$

Therefore, $H^{s_0}(C) > 0$.   □

§8.3. Generalized Cantor Sets—Part I

Our definition of the Cantor ternary set did not rely on the ternary representation of the points in the interval $[0, 1]$. We introduced this representation
after the set was defined in an effort to better understand the nature of the set.
However, we could have used such a representation to define the Cantor set.
Indeed, Gilman [G1] has defined a general class of sets with the properties
of the Cantor set and a corresponding class of functions which have the
properties of ω using bases other than 2 and 3.

Let α be any positive integer greater than 2. Then any number $x \in [0, 1]$
can be expressed in base α, i.e.,

$$(8.3.1) \qquad x = \frac{a_1}{\alpha} + \frac{a_2}{\alpha^2} + \cdots + \frac{a_n}{\alpha^n} + \cdots = {}_{.\alpha}a_1 a_2 \ldots a_n \ldots,$$

where $a_1, a_2, \ldots, a_n, \ldots$ are integers such that $0 \leq a_i < \alpha$, $i = 1, 2, \ldots$. Let
$\alpha - 1$ be represented as the product of two integers as follows:

$$\alpha - 1 = q(\beta - 1), \quad q \geq 2, \beta \geq 2.$$

Let $P_{\alpha\beta}$ be the set of all $x \in [0, 1]$ for which the a_i's in the representation (8.3.1)
of x are integral multiples of q. Then $P_{\alpha\beta}^c$ contains those $x \in [0, 1]$ for which
no representation in the base α has every digit as a multiple of q. Thus, if we
take $\alpha = 3$ and $\beta = 2$, then $P_{32} = C$, the Cantor ternary set.

Associated with $P_{\alpha\beta}$, we define a function $\omega_{\alpha\beta}$. Let $x = {}_{.\alpha}a_1 a_2 \ldots = a_1/\alpha + a_2/\alpha^2 + \cdots$. When $a_i = 0 \pmod{q}$ and $a_i = qb_i$, we define

$$\omega_{\alpha\beta}(x) = \frac{b_1}{\beta} + \frac{b_2}{\beta^2} + \cdots + \frac{b_n}{\beta^n} + \cdots.$$

When $a_i = qb_i$ $(i = 1, 2, \ldots, n - 1)$ but $a_n \neq 0 \pmod{q}$, we define

$$\omega_{\alpha\beta}(x) = \frac{b_1}{\beta} + \frac{b_2}{\beta^2} + \cdots + \frac{b_{n-1}}{\beta^{n-1}} + \frac{b_n}{\beta^n},$$

where $b_n = [a_n/q] + 1$.

We leave it to the reader to show that the sets $P_{\alpha\beta}$ and their associated
functions $\omega_{\alpha\beta}$ share many of the same properties as the Cantor ternary set
and its associated function (Exercises 1–11).

§8.4. Generalized Cantor Sets—Part II

Herzog and Bissinger [HB] have constructed another class of sets of the Cantor type based on the representation of numbers as simple continued fractions. Throughout this section, we shall use the letter c (with or without a subscript) to denote a positive integer.

Definition 8.4.1. Let $0 < x < 1$. If the positive integers c_1, c_2, c_3, ... are successively the denominators (partial quotients) appearing in the expansion of x as a simple continued fraction, then we will write

$$x = \cfrac{1}{c_1 + \cfrac{1}{c_2 + \cfrac{1}{c_3 + \cdots}}} = \{c_1, c_2, c_3, \ldots\}.$$

If $x \in (0, 1)$ is rational, there are two such expansions, namely

$$x = \{c_1, \ldots, c_{n-1}, c_n\} = \{c_1, \ldots, c_{n-1}, c_n - 1, 1\},$$

where $c_n \geq 2$. We will allow either expansion to be used for such an x. Also, $1 = \{1\}$.

Definition 8.4.2. We define the set E as the set consisting of 0 and all $x \in (0, 1)$ which can be expanded in a simple continued fraction, none of whose denominators equals 1; that is, E contains 0 and all finite and simple continued fractions of the form $x = \{c_1 + 1, c_2 + 1, \ldots\}$. Note that $x = \{c_1 + 1, \ldots, c_n + 1\}$ does belong to E although it can also be written as $\{c_1 + 1, \ldots, c_n, 1\}$.

The complement of E (relative to the interval $[0, 1]$) will be denoted by D. Thus, D contains 1 and all open intervals with endpoints ξ and ξ' where

(8.4.3)
$$\xi = \{c_1 + 1, c_2 + 1, \ldots, c_{n-1} + 1, 1\}$$
$$= \{c_1 + 1, c_2 + 1, \ldots, c_{n-1} + 2\},$$
$$\xi' = \{c_1 + 1, c_2 + 1, \ldots, c_{n-1} + 1, 2\}.$$

Such an interval will be called a D_n-*interval*. We list here some of the D_n-intervals.

$$n = 1: \quad (\{2\}, \{1\}) = \left(\frac{1}{2}, 1\right)$$

$$n = 2: \quad (\{3\}, \{2, 2\}) = \left(\frac{1}{3}, \frac{2}{5}\right)$$

$$(\{4\}, \{3, 2\}) = \left(\frac{1}{4}, \frac{2}{7}\right)$$

$$(\{5\}, \{4, 2\}) = \left(\frac{1}{5}, \frac{2}{9}\right)$$

$$n = 3: \quad (\{2,2,2\},\{2,3\}) = \left(\frac{5}{12},\frac{3}{7}\right)$$

$$(\{2,3,2\},\{2,4\}) = \left(\frac{7}{16},\frac{4}{9}\right)$$

$$(\{3,2,2\},\{3,3\}) = \left(\frac{5}{17},\frac{3}{10}\right)$$

$$n = 4: \quad (\{2,2,3\},\{2,2,2,2\}) = \left(\frac{7}{17},\frac{12}{29}\right)$$

There is one D_1-interval but infinitely many D_n-intervals for each $n \geq 2$. Note than any two D_n-intervals (for equal or unequal values of n) do not overlap and do not even have an endpoint in common.

The points of E (except for 0) can be divided into two classes: E_r which contains all the rational numbers in E (except 0) and E_i which contain the irrational numbers in E. Then $x \in E_r$ if and only if x is the endpoint of a D_n-interval (with the exception of 1, which is not in E at all); 0 will be considered to belong to neither E_r or E_i.

We leave it to the reader to show that E is closed, perfect, uncountable, nowhere dense, and has Lebesgue measure zero.

Comparing the set E to the set $P_{\alpha\beta}$ of §8.3, we see that the set E_r corresponds to the set $P_{\alpha\beta}^+ \cup P_{\alpha\beta}^-$ and the set E_i corresponds to the set $P_{\alpha\beta}^0$ (see Exercise 11). The set $P_{\alpha\beta}$ is symmetric about the point $x = \frac{1}{2}$; the set E has no such symmetry. In the construction of the Cantor ternary set, the intervals E_{ni} for a fixed n are finite in number and of equal length. However, in the construction of E, for $n \geq 2$, there are infinitely many D_n-intervals of unequal length.

We now define the Cantor-type function ϕ associated with the set E. We first give the definition for $x \in E$:

$\phi(0) = 0$,
$\phi(x) = \{c_1, c_2, \ldots, c_n\}$, $x \in E_r$ and $x = \{c_1 + 1, c_2 + 1, \ldots, c_n + 1\}$,
$\phi(x) = \{c_1, c_2, \ldots\}$, $x \in E_i$ and $x = \{c_1 + 1, c_2 + 1, \ldots\}$.

Additionally, we define $\phi(1) = 1$ and show that the values of ϕ at both endpoints of each D_n-interval are equal. For the D_1-interval $(\frac{1}{2}, 1)$, we have $\phi(\frac{1}{2}) = \phi(\{2\}) = \{1\} = 1 = \phi(1)$. For a D_n-interval with $n \geq 2$, we see that if ξ and ξ' are given by (8.4.3), we have

$$\phi(\xi) = \{c_1, c_2, \ldots, c_{n-1} + 1\} = \{c_1, c_2, \ldots, c_{n-1}, 1\} = \phi(\xi').$$

We now define ϕ for any interior point of a D_n-interval as the common value of ϕ at the endpoints of that interval. We list the values of ϕ for the D_n-

intervals given previously.

$$\phi(x) = \begin{cases} \{1\} = 1, & x \in \left[\dfrac{1}{2}, 1\right] \\[2mm] \{2\} = \dfrac{1}{2}, & x \in \left[\dfrac{1}{3}, \dfrac{2}{5}\right] \\[2mm] \{3\} = \dfrac{1}{3}, & x \in \left[\dfrac{1}{4}, \dfrac{2}{7}\right] \\[2mm] \{4\} = \dfrac{1}{4}, & x \in \left[\dfrac{1}{5}, \dfrac{2}{9}\right] \\[2mm] \{1,2\} = \dfrac{2}{3}, & x \in \left[\dfrac{5}{12}, \dfrac{3}{7}\right] \\[2mm] \{1,3\} = \dfrac{3}{4}, & x \in \left[\dfrac{7}{16}, \dfrac{4}{9}\right] \\[2mm] \{2,2\} = \dfrac{2}{5}, & x \in \left[\dfrac{5}{17}, \dfrac{3}{10}\right] \\[2mm] \{1,1,2\} = \dfrac{3}{5}, & x \in \left[\dfrac{7}{17}, \dfrac{12}{29}\right]. \end{cases}$$

In the exercises, you will be asked to prove that ϕ has the properties of the functions $\omega_{\alpha\beta}$ of §8.2. However, it will be seen (Exercise 15) that the curve $y = \phi(x)$ has all rational numbers $y \in (0, 1]$ as ordinates of its horizontal segments, whereas the functions $\omega_{\alpha\beta}$ have only certain rational numbers as such ordinates, namely those of the form $y = m/\beta^r$, r and m being positive integers.

In addition to the properties of ϕ given by the exercises, in [HB] it is stated without proof that ϕ satisfies a Lipschitz condition of order $\alpha = (\log b)/(\log a) = 0.546$, where

$$a = \{2, 2, \ldots\} = 2^{1/2} - 1$$

and

$$b = \phi(a) = \{1, 1, \ldots\} = \frac{5^{1/2} - 1}{2},$$

but that ϕ does not satisfy a Lipschitz condition of any higher order. When $\alpha = (\log b)/(\log a)$, the "best" value for the Lipschitz constant is $10^{\alpha}/2 = 1.758$.

§8.5. Cantor-like Sets

In this section, we generalize the method of construction of the Cantor ternary set of Example 1.3.3 to obtain a class of sets which share most of the properties of the Cantor ternary set.

We begin with the interval $[0,1]$ and proceed inductively. Remove an open interval $I_{1,1}$ from $[0,1]$ with center at $\frac{1}{2}$ and length less than 1. This leaves two residual intervals $J_{1,1}$ and $J_{1,2}$, each having length $<\frac{1}{2}$. Suppose that the nth step has been completed, leaving closed intervals $J_{n,1}, \ldots, J_{n,2^n}$, each having length $<1/2^n$. Then for the $(n+1)$st step, we remove from each interval $J_{n,k}$ an open interval $I_{n+1,k}$ concentric with $J_{n,k}$ and of length $<1/2^n$. Let

$$P_n = \bigcup_{k=1}^{2^n} J_{n,k} \quad \text{and} \quad P = \bigcap_{n=1}^{\infty} P_n.$$

We will call P a *Cantor-like set*.

We leave it to the reader to verify that P is closed, nowhere dense, and perfect [see D1]. Unlike the Cantor ternary set and the generalized Cantor sets of §§8.3 and 8.4, the Cantor-like sets do not in general have Lebesgue measure zero (see Example 8.5.4).

A particular case of this type of construction is when $|J_{11}| = |J_{12}| = \eta < \frac{1}{2}$, $|J_{21}| = |J_{22}| = |J_{23}| = |J_{24}| = \eta^2$, and so forth. Note for this case that at each stage the J intervals are divided in the same proportion as the orginal interval $[0,1]$. We obtain the Cantor ternary set if $\eta = \frac{1}{3}$.

Theorem 8.5.1 [D1, p. 50]. *Let P_η denote the Cantor-like set obtained by the previous construction. Then the Hausdorff dimension of P_η is $-(\log 2)/(\log \eta)$.*

Proof. The proof is the same as that of Theorem 8.2.22, substituting η for $\frac{1}{3}$. $\qquad\square$

Corollary 8.5.2 [D1, p. 511]. *For each $\alpha \in [0,1]$, there exists a set $A \subset \mathbf{R}^1$ with Hausdorff dimension α.*

Proof. The case $\alpha = 1$ is covered by Theorem 8.2.10. For $0 < \alpha < 1$, take $\eta = \exp[-(\log 2)/\alpha]$ in the previous theorem. For $\alpha = 0$, take $A = \{x\}$. $\qquad\square$

EXAMPLE 8.5.3 [D1, p. 26]. Consider the special case of Cantor-like sets where for some fixed $\alpha \in (0,1]$, the removed open intervals, denoted by $I_{n,k}^\alpha$, are such that $|I_{n,k}^\alpha| = \alpha/3^n$ for $k = 1, \ldots, 2^{n-1}$ and for each n. Denote the set obtained by this construction as P^α. If $\alpha = 1$, we obtain the Cantor ternary set C. Let g_n be the increasing piecewise linear function from $[0,1] \to [0,1]$ which maps the endpoints of $J_{n,k}^\alpha$ onto those of $J_{n,k}$ for $k = 1, \ldots, 2^n$. Then for $n > m$, g_n and g_m differ only on $J_{m,k}^\alpha$ for $k = 1, \ldots, 2^m$ and $|g_m - g_n| < |J_{n,k}| = 1/3^m$. Now $\{g_m(x)\}$, for each x, is convergent; let g be the limit function. Also, $|g_m(x) - g(x)| = \lim_n |g_m(x) - g_n(x)| \leq 1/3^m$. Hence, g_n is uniformly convergent to g. Thus, g is continuous and increasing. Also, $g([0,1]) = [0,1]$ and $g(I_{n,k}^\alpha) = I_{n,k}$ for each n and for each k. Thus, $g(P^\alpha) = C$. We now show that g is one-to-one. Let $(x,y) \subset [0,1]$. If either x or y is in a removed interval, then $g(y) > g(x)$. If $x, y \in P^\alpha$, using the fact that P^α is nowhere dense, there exists

$I^\alpha_{n,k} \subset (x, y)$ for some n and k. Then $g(y) - g(x) \geq 1/3^n$. Thus, g is strictly increasing and hence one-to-one.

Another way [D1, p. 24] to apply this construction method for Cantor-like sets is to let $|J_{11}| = |J_{12}| = \xi_1 < \frac{1}{2}$, $|J_{21}| = |J_{22}| = |J_{23}| = |J_{24}| = \xi_1\xi_2$, and so on, where $\xi_{n+1} < \xi_n/2$ for each n. In this case, at each step the J intervals are equal in length but the proportions are allowed to change from step to step.

EXAMPLE 8.5.4 [J2, p. 83]. Start with the closed interval $[0, 1]$. Choose any sequence of positive numbers $\{\eta_k\}$ such that

$$1 = \eta_0 > 2\eta_1 > 4\eta_2 > \cdots > 2^k\eta_k > \cdots.$$

Now remove from $[0, 1]$ the open interval $I_{1,1}$ with center $\frac{1}{2}$ and length $1 - 2\eta_1$, leaving two closed intervals $J_{1,1}$ and $J_{1,2}$, each of length η_1. Now from each of the closed intervals $J_{1,1}$ and $J_{1,2}$, remove concentric open intervals, $I_{1,1}$ and $I_{1,2}$, respectively, with length $\eta_1 - 2\eta_2$, leaving four closed intervals $J_{2,1}$, $J_{2,2}$, $J_{2,3}$, and $J_{2,4}$ each of length η_2. Continue this process indefinitely; at the nth step, there remain 2^n closed disjoint intervals each of length η_n. Let P be the Cantor-like set obtained from the intersection of all the unions of the closed intervals $J_{n,k}$. Then

$$mP = \lim_{n \to \infty} 2^n\eta_n.$$

If $\eta_n = 3^{-n}$, we obtain the Cantor ternary set. In spite of the fact that P contains no intervals, we can make mP as close to 1 as we choose. Given any $\theta \in [0, 1)$, choose

$$2^n\eta_n = \frac{\theta n + 1}{n + 1}.$$

Then $mP = \theta$.

Still another way [D1, p. 24] to vary this construction is to choose the removed open intervals nonconcentric with the closed intervals from which they are being removed. The centers of the open intervals may be chosen to be a fixed convex combination of the endpoints of the closed intervals.

We now construct functions corresponding to the Cantor ternary function for the Cantor-like sets of Example 8.5.4 [J2, pp. 86–89]. In the first stage of that construction, we removed the open interval $I_{1,1}$ with center $\frac{1}{2}$. It will make our presentation somewhat simpler if we rename the interval $I_{1,1}$ as $I_{1/2}$. At the second stage, we removed the two open intervals $I_{2,1}$ and $I_{2,2}$ having centers $\frac{1}{4}$ and $\frac{3}{4}$, respectively. We now rename these intervals $I_{1/4}$ and $I_{3/4}$. At the third stage, the open intervals removed will be denoted $I_{1/8}$, $I_{3/8}$, and so on. In addition, let $I_0 = (-\infty, 0)$ and $I_1 = (1, \infty)$.

Let D denote the dyadic rationals in $[0, 1]$; i.e., numbers of the form $m/2^n$ for integers m and n such that $0 \leq m \leq 2^n$, $0 \leq n$. Then, for each $r \in D$, the method of construction of Example 8.5.4 produces an open interval I_r with center r. Note that if $r < r'$, then I_r is entirely to the left of $I_{r'}$.

Let us first define a function f on the union of the open intervals I_r by

$$f(x) = r \quad \text{for all } x \in I_r, r \in D.$$

Since $P = R^1 - \bigcup_{r \in D} I_r$, f is now defined on P^c. We leave it to the reader to confirm that f is an increasing function which is constant on all the intervals I_r and that f is uniformly continuous on P^c. Recall that a function F which is uniformly continuous on a subset $E \subset R^1$ can be extended to a unique function which is continuous on $\bar{E}$ and which agrees with F on E. Since P is nowhere dense, $\bar{P^c} = R^1$. Thus, there exists a unique continuous function on R^1 which agrees with f on P^c; we will denote the extension by f also. This function is often called the *Lebesgue function* for the Cantor-like set P and has the following properties:

(8.5.5) $f = r$ on each open interval I_r for any $r \in D$,

(8.5.6) f is increasing,

(8.5.7) $f : R^1 \to [0, 1]$,

(8.5.8) f is continuous; in fact, $|x - y| \le \eta_k \Rightarrow |f(x) - f(y)| \le 2^{-k}$.

Lemma 8.5.9 [J2, p. 90]. *Let r be a dyadic rational in $[0, 1)$ and suppose r is given by*

$$r = \sum_{j=1}^{k} \varepsilon_j 2^{-j},$$

where $\varepsilon_j = 0$ or 1. Then the corresponding deleted open interval $I_r = (a, b)$ has a right endpoint given by

$$b = \sum_{j=1}^{k} \varepsilon_j (\eta_{j-1} - \eta_j).$$

Proof. The proof is by induction. For $k = 1$, we have $r = \varepsilon_1 2^{-1}$. If $\varepsilon_1 = 0$, then $I_0 = (-\infty, 0)$ and the formula gives $b = 0$. If $\varepsilon_1 = 1$, then $I_{1/2} = (\eta_1, \eta_0 - \eta_1)$ and the formula gives $b = \eta_0 - \eta_1$. Thus, the formula is correct for $k = 1$. Now we assume it is correct for $k - 1$ and prove that it holds for k; that is, assume

$$r = \sum_{j=1}^{k} \varepsilon_j 2^{-j}$$

and $I_r = (a, b)$. We can assume that $\varepsilon_k = 1$; otherwise, the result for k would hold by the assumption that it holds for $k - 1$. The open interval I_r is deleted from a closed interval of length η_{k-1}. Since

$$r - \frac{1}{2^k} = \sum_{j=1}^{k-1} \varepsilon_j 2^{-j},$$

the induction hypothesis gives the right endpoint of $I_{r-1/2^k}$ is

$$\sum_{j=1}^{k-1} \varepsilon_j(\eta_{j-1} - \eta_j).$$

Thus,

$$b = \sum_{j=1}^{k-1} \varepsilon_j(\eta_{j-1} - \eta_j) + (\eta_{k-1} - \eta_k). \qquad \square$$

Theorem 8.5.10 [J2, p. 92]. *Let P be a Cantor-like set as constructed in Example 8.5.4. Let $x \in [0,1]$. Then $x \in P \Leftrightarrow$*

$$x = \sum_{j=1}^{\infty} \varepsilon_j(\eta_{j-1} - \eta_j),$$

where $\varepsilon_j = 0$ or 1. This representation of x is unique. Furthermore, if f is the Lebesgue function for P, then

$$f(x) = \sum_{j=1}^{\infty} \varepsilon_j 2^{-j}.$$

Proof. First, we suppose that x has such a representation. Then define

$$x_k = \sum_{j=1}^{k} \varepsilon_j(\eta_{j-1} - \eta_j).$$

By the previous lemma, x_k is the right endpoint of a deleted interval I_{r_k}; therefore, $x_k \in P$. Since $x_k \to x$ and P is closed, $x \in P$. Also,

$$f(x_k) = r_k = \sum_{j=1}^{k} \varepsilon_j 2^{-j}.$$

Because f is continuous, we obtain the desired result for $f(x)$ by letting $k \to \infty$.

Conversely, assume $x \in P$. Then for any k, we know that x belongs to one of the closed J_r intervals of length η_k. Now r has the form

$$r = \sum_{j=1}^{k} \varepsilon_j 2^{-j}.$$

The previous lemma gives the right endpoints of I_r. Therefore,

$$\sum_{j=1}^{k} \varepsilon_j(\eta_{j-1} - \eta_j) \le x \le \sum_{j=1}^{k} \varepsilon_j(\eta_{j-1} - \eta_j) + \eta_k,$$

for unique $\varepsilon_1, \ldots, \varepsilon_k$. At the $(k+1)$st stage of the construction of P, we obtain a similar inequality

$$\sum_{j=1}^{k+1} \delta_j(\eta_{j-1} - \eta_j) \le x \le \sum_{j=1}^{k+1} \delta_j(\eta_{j-1} - \eta_j) + \eta_{k+1}.$$

But $\delta_j = \varepsilon_j$ for $1 \le j \le k$. Since $0 \le \delta_{k+1} \le 1$, the previous inequality implies

$$\sum_{j=1}^{k} \delta_j(\eta_{j-1} - \eta_j) \le x \le \sum_{j=1}^{k} \delta_j(\eta_{j-1} - \eta_j) + \eta_k.$$

But the uniqueness at the kth stage means that $\delta_j = \varepsilon_j$ for $1 \le j \le k$. Because $\eta_k \to 0$ as $k \to \infty$, we conclude that

$$x = \sum_{j=1}^{\infty} \varepsilon_j(\eta_{j-1} - \eta_j).$$

Is this representation unique? Suppose not; then we would have

$$\sum_{j=1}^{\infty} \varepsilon_j(\eta_{j-1} - \eta_j) = x = \sum_{j=1}^{\infty} \delta_j(\eta_{j-1} - \eta_j).$$

There would be a smallest $k \ge 1$ such that $\varepsilon_k \ne \delta_k$. Suppose that for this k, $\varepsilon_k = 1$ and $\delta_k = 0$. Then

$$\eta_{k-1} - \eta_k \le \sum_{j=k}^{\infty} \varepsilon_j(\eta_{j-1} - \eta_j)$$

$$= \sum_{j=k}^{\infty} \delta_j(\eta_{j-1} - \eta_j)$$

$$\le \sum_{j=k+1}^{\infty} (\eta_{j-1} - \eta_j) = \eta_k.$$

Thus, $\eta_{k-1} \le 2\eta_k$, which is a contradiction.   □

In Theorem 8.1.21, we showed that the Cantor ternary function satisfied a Lipschitz condition of order $\alpha = (\log 2)/(\log 3)$. We now generalize this result for the Cantor-like sets of Example 8.5.4 [J2, p. 95].

Let h be a function defined on $[0, 1]$ such that

h is continuous and increasing,
$h(0) = 0$, $h(1) = 1$, and
h is strictly concave.

The third condition means that any chord connecting two points of the graph of $y = h(t)$ lies strictly below the graph; that is, if $0 \le a < b < c \le 1$, then

(8.5.11) $$h(b) > \frac{c - b}{c - a} h(a) + \frac{b - a}{c - a} h(c).$$

An example of a concave function h is one which is of class C^2 such that $h'' \le 0$; if $h \in C^2$ and $h'' < 0$, then such an h is strictly concave. For $0 < \theta < 1$, $h(t) = t^\theta$ is strictly concave. Another such function is $h(t) = mt - (m - 1)t^2$ for any $m > 1$.

By letting $a = 0$, $b = t$, and $c = 2t$ in (8.5.11), we see that

$$2h(t) > h(2t) \quad \text{if } 0 \le t \le \tfrac{1}{2}.$$

We will now use a fixed concave function h to choose the sequence $\{\eta_k\}$ in the construction of the Cantor-like set of Example 8.5.4. We will choose η_k such that $h(\eta_k) = 2^{-k}$, $k = 0, 1, 2, \ldots$. The previous inequality then gives

$$h(\eta_{k-1}) = 2^{1-k} = 2h(\eta_k) > h(2\eta_k).$$

Since h is increasing, $\eta_{k-1} > 2\eta_k$. Since $h(\eta_0) = 1$, we have $\eta_0 = 1$; so that the sequence $\{\eta_k\}$ satisfies the necessary requirements. Then the measure of the set P constructed using this sequence is

$$mP = \lim_{k \to \infty} 2^k \eta_k = \lim_{k \to \infty} \frac{\eta_k}{h(\eta_k)}.$$

Since h is concave, $h'(0)$ exists. Because $\eta_k \to 0$,

$$mP = \frac{1}{h'(0)},$$

where this is considered to be zero if $h'(0) = \infty$. If $h(t) = t^\theta$ for $\theta \in (0, 1)$, then $mP = 0$. If $h(t) = mt - (m - 1)t^2$ for $m \in (1, \infty)$, then $mP = 1/m$.

Theorem 8.5.12 [J2, p. 97]. *For the Cantor-like set P constructed by means of a fixed concave function h as described above, the associated Lebesgue function f satisfies the inequality*

$$|f(x) - f(y)| \le h(|x - y|) \quad \text{for all } x, y \in [0, 1].$$

Proof. Since the union of the removed open intervals I_r for $r \in D$ is dense in $[0, 1]$ and since f and h are continuous, it suffices to prove the result for x and y in the union of the I_r's. Since f is constant on I_r, we will assume that x and y belong to two different I_r's. By symmetry, we can assume that $x < y$, that is,

$$x \in I_{r_1}, \quad y \in I_{r_2}, \quad 0 < r_1 < r_2 < 1.$$

Then we need to show that

$$r_2 - r_1 \le h(y - x).$$

Let $I_{r_i} = (a_i, b_i)$, $i = 1, 2$. Then $a_1 < x < b_1$ and $a_2 < y < b_2$. Thus, $y - x > a_2 - b_1 > 0$. Since h is increasing, it will be sufficient to show that

$$r_2 - r_1 \le h(a_2 - b_1).$$

By Exercise 22, we have $J_{1-r_2} = (1 - b_2, 1 - a_2)$. By Lemma 8.5.9, if

$$r_1 = \sum_{j=1}^{k} \varepsilon_j 2^{-j} \quad \text{and} \quad 1 - r_2 = \sum_{j=1}^{k} \delta_k 2^{-j},$$

then

$$b_1 = \sum_{j=1}^{k} \varepsilon_j(\eta_{j-1} - \eta_j) \quad \text{and} \quad 1 - a_2 = \sum_{j=1}^{k} \delta_j(\eta_{j-1} - \eta_j).$$

Each ε_j and each δ_j is 0 or 1. Thus, we have to prove that

$$1 - \sum_{j=1}^{k} (\varepsilon_j + \delta_j) 2^{-j} \leq h\left(1 - \sum_{j=1}^{k} (\varepsilon_j + \delta_j)(\eta_{j-1} - \eta_j)\right),$$

provided the left side of the above inequality is positive. Since ε_j and δ_j are 0 or 1, we let $\alpha_j = \varepsilon_j + \delta_j$, where α_j is 0, 1, or 2. Therefore, we have to prove

(8.5.13) $$1 - \sum_{j=1}^{k} \alpha_j 2^{-j} \leq h\left(1 - \sum_{j=1}^{k} \alpha_j(\eta_{j-1} - \eta_j)\right),$$

if the left side is positive and α_j is 0, 1 or 2. We first require a lemma.

Lemma 8.5.14 [J2, p. 99]. *If α_j is 0, 1, or 2 and*

$$\sum_{j=1}^{k} \alpha_j 2^{-j} < 1,$$

then

$$\sum_{j=1}^{k} \alpha_j(\eta_{j-1} - \eta_j) \leq 1 - \eta_k.$$

Proof. The proof is by induction; the case $k = 1$ is obvious. Assume $k > 1$ and

$$\sum_{j=1}^{k} \alpha_j 2^{-j} < 1.$$

Note that this requires α_1 to be 0 or 1. If $\alpha_1 = 0$, then

$$\sum_{j=1}^{k} \alpha_j(\eta_{j-1} - \eta_j) \leq \sum_{j=2}^{k} 2(\eta_{j-1} - \eta_j) = 2(\eta_1 - \eta_k) < 2\eta_1 - \eta_k < 1 - \eta_k,$$

so the conclusion holds. If $\alpha_1 = 1$, then rewrite the hypothesis in the form

$$\sum_{j=2}^{k} \alpha_j 2^{-j} < \tfrac{1}{2}.$$

Multiply by 2 to get

$$\sum_{j=1}^{k-1} \alpha_{j+1} 2^{-j} < 1.$$

Define a new function

$$h^*(x) = \frac{h(\eta_1 x)}{h(\eta_1)} = 2h(\eta_1 x), \quad 0 \leq x \leq 1.$$

Then h^* is a strictly concave function and the corresponding numbers η_j^* are given by

$$h^*(\eta_j^*) = 2^{-j}.$$

Therefore, $\eta_1 \eta_j^* = \eta_{j+1}$. By the induction hypothesis, we have

$$\sum_{j=1}^{k-1} \alpha_{j+1}(\eta_{j-1}^* - \eta_j^*) \le 1 - \eta_{k-1}^*;$$

that is,

$$\sum_{j=1}^{k-1} \alpha_{j+1}(\eta_j - \eta_{j+1}) \le \eta_1 - \eta_k.$$

Since $\alpha_1 = 1$, this implies that

$$\sum_{j=1}^{k} \alpha_j(\eta_{j-1} - \eta_j) \le 1 - \eta_1 + \eta_1 - \eta_k = 1 - \eta_k,$$

so the conclusion of the lemma is valid. $\qquad\square$

Now for the proof of (8.5.13); we shall prove it by induction on the number

$$\sum_{j=1}^{k} \alpha_j.$$

If this sum is 0, then all $\alpha_j = 0$ and (8.5.13) follows. Therefore, we assume $\alpha_k \ne 0$. The induction hypothesis yields

$$1 - \sum_{j=1}^{k} \alpha_j 2^{-j} = \left(1 - \sum_{j=1}^{k-1} \alpha_j 2^{-j} - (\alpha_k - 1)2^{-k}\right) - 2^{-k} \le h(x) - 2^{-k},$$

where

$$x = 1 - \sum_{j=1}^{k-1} \alpha_j(\eta_{j-1} - \eta_j) - (\alpha_k - 1)(\eta_{k-1} - \eta_k).$$

Inequality (8.5.13) will follow if we show that

$$h(x) - 2^{-k} \le h(x - \eta_{k-1} + \eta_k).$$

We will justify the last inequality by showing that

$$h(x) - h(x - \eta_{k-1} + \eta_k) \le h(\eta_{k-1}) - h(\eta_k).$$

Since h is concave, its difference quotients decrease as the chords move to the right (see §3.7). Thus, it suffices to prove that $x \ge \eta_{k-1}$. By definition of x, this means that

$$\sum_{j=1}^{k} \alpha_j(\eta_{j-1} - \eta_j) \le 1 - \eta_k.$$

But this result holds by the previous lemma if

$$\sum_{j=1}^{k} \alpha_j 2^{-j} < 1$$

which holds if the left side of (8.5.13) is positive. $\qquad\square$

§8.6. Strictly Increasing Singular Functions

The function constructed in the preceding sections of this chapter, based on the generalized Cantor sets, are singular functions; that is, continuous functions whose derivatives are zero almost everywhere. These generalized Cantor functions are also increasing, but not strictly so. We now consider some examples of singular functions which are strictly increasing.

EXAMPLE 8.6.1 [S2]. Let us consider two points in the plane, $P = (x, y)$ and $Q = (x + \Delta x, y + \Delta y)$, with $\Delta x > 0$ and $\Delta y > 0$. Let λ_0 and λ_1 be two positive numbers such that $\lambda_0 \neq \lambda_1$ and $\lambda_0 + \lambda_1 = 1$. Let R be the point whose coordinates are $(x + \Delta x/2, y + \lambda_0 \Delta y)$. Thus, the horizontal distance between P and R, or between Q and R, is $\Delta x/2$, whereas the vertical distance between P and R is $\lambda_0 \Delta y$ and the vertical distance between Q and R is $(1 - \lambda_0)\Delta y = \lambda_1 \Delta y$. If we replace the straight line PQ by the broken line PRQ, we will say that we have performed the transformation $T(\lambda_0, \lambda_0)$ on PQ.

For $x \in [0, 1]$, define $f_0(x) = x$. Thus, the graph of $f_0(x)$ is the straight line connecting the origin $O = (0, 0)$ to the point $A = (1, 1)$. Perform on OA the transformation $T(\lambda_0, \lambda_1)$ to get a broken line consisting of two straight line segments. This broken line represents an increasing function which we shall call f_1. Then we perform on each of these two segments the transformation $T(\lambda_0, \lambda_1)$ to get a broken line consisting of $4 = 2^2$ straight line segments and representing an increasing function f_2. Proceeding in this manner, at the pth step, we have a strictly increasing function f_p, $f_p(0) = 0$ and $f_p(1) = 1$, whose graph is a polygonal line consisting of 2^p straight line segments, the vertices having x coordinates $k/2^p$ $(k = 1, 2, \ldots, 2^p - 1)$.

Let $\mu = \max(\lambda_0, \lambda_1)$. Then $\mu < 1$ and

$$|f_{p+1} - f_p| \leq \mu^p.$$

Thus, f_p converges uniformly to a continuous function f, $f(0) = 0$ and $f(1) = 1$. This function f is strictly increasing; for every p the vertices of the curve $y = f_p(x)$ belong to the curve $y = f(x)$. If f were constant in some interval, there would be an index p for which two different vertices of $y = f_p(x)$ would have the same ordinate, which is not possible.

The ordinate of the vertex of $y = f_p(x)$ whose abscissa is given by

$$\frac{\theta_1}{2} + \frac{\theta_2}{2^2} + \cdots + \frac{\theta_p}{2^p} \quad (\theta_i = 0 \text{ or } 1)$$

is

$$\lambda_0[\theta_1 + \lambda_{\theta_1}\theta_2 + \lambda_{\theta_1}\lambda_{\theta_2}\theta_3 + \cdots + \lambda_{\theta_1}\lambda_{\theta_2}\cdots\lambda_{\theta_{p-1}}\theta_p].$$

Thus, by continuity of f, if

(8.6.2) $$x = \frac{\theta_1}{2} + \frac{\theta_2}{2^2} + \cdots + \frac{\theta_p}{2^p} + \cdots,$$

then

(8.6.3) $f(x) = \lambda_0[\theta_1 + \lambda_{\theta_1}\theta_2 + \lambda_{\theta_1}\lambda_{\theta_2}\theta_3 + \cdots + \lambda_{\theta_1}\lambda_{\theta_2}\cdots\lambda_{\theta_{p-1}}\theta_p + \cdots].$

This series is obviously convergent. If x has two different dyadic representations, the formula in (8.6.3) yields the same value for $f(x)$.

Note that if x and $x' > x$ have the first p digits of their dyadic representations identical and equal to $\theta_1, \theta_2, \ldots, \theta_p$, then

(8.6.4) $f(x') - f(x) < \lambda_{\theta_1}\lambda_{\theta_2}\cdots\lambda_{\theta_p}.$

We now show that f is a singular function. It is known [F1, p. 195] that almost all numbers in $(0, 1)$ are "normal" in the scale of 2, that is, are such that

$$\theta_1 + \theta_2 + \cdots + \theta_p = \frac{p}{2} + o(p) \quad \text{when } p \to \infty.$$

Let N be the set of these normal numbers. Thus, $mN = 1$. Let x be a fixed number in N given by (8.6.2) and p a positive integer. Let

$$\varepsilon_{p+1} = \begin{cases} 1 & \text{if } \theta_{p+1} = 0 \\ -1 & \text{if } \theta_{p+1} = 1. \end{cases}$$

Then the number $x + \varepsilon_{p+1}/2^{p+1}$ has a dyadic representation whose first p digits are the same as those of x. Thus, by (8.6.4)

$$\left| f\left(x + \frac{\varepsilon_{p+1}}{2^{p+1}}\right) - f(x) \right| < \lambda_{\theta_1}\lambda_{\theta_2}\cdots\lambda_{\theta_p}.$$

Since $x \in N$,

$$\theta_1 + \theta_2 + \cdots + \theta_p = \frac{p}{2} + \phi(p)$$

with $|\phi(p)|/p \to 0$ when $p \to \infty$. Hence,

$$\lambda_{\theta_1}\lambda_{\theta_2}\cdots\lambda_{\theta_p} = \lambda_0^{p/2-\phi(p)}\lambda_1^{p/2+\phi(p)} < (\lambda_0\lambda_1)^{p/2-|\phi(p)|}.$$

Therefore,

(8.6.5) $2^{p+1}\left| f\left(x + \frac{\varepsilon_{p+1}}{2^{p+1}}\right) - f(x) \right| < (2(\lambda_0\lambda_1)^{1/2})^p \frac{2}{(\lambda_0\lambda_1)^{|\phi(p)|}}.$

Because $\lambda_0 \neq \lambda_1$ and $\lambda_0 + \lambda_1 = 1$, we have

$$2(\lambda_0\lambda_1)^{1/2} < 1.$$

Thus, the second term on the right side of (8.6.5) tends to 0 as $p \to \infty$. Hence, if f has a derivative at the point x, this derivative must be 0. But, since f is increasing, f' exists and is finite almost everywhere. Hence, $f' = 0$ a.e.; thus, f is singular.

Now we will show that f satisfies a Lipschitz condition of order $|\log \mu|/(\log 2)$. If $k/2^p$ and $(k + 1)/2^p$ are the x coordinates of two vertices of f_p, then

$$f_p\left(\frac{k+1}{2^p}\right) - f_p\left(\frac{k}{2^p}\right) < \mu^p,$$

from which it follows that if $1/2^{p+1} \leq x' - x < 1/2^p$, then

$$f(x') - f(x) < 2\mu^p \leq 2\mu^{-1}\mu^{\log(x'-x)^{-1}/(\log 2)} = 2\mu^{-1}(x' - x)^{|\log \mu|/(\log 2)}.$$

In the construction of the preceding example, we performed an identical transformation at each step on each of the polygonal line segments from the preceding step. It is possible to generalize this method of construction by using a different transformation at each step. As before, we begin with the function $F_0(x) = x$ for $x \in [0, 1]$ and let $O = (0, 0)$ and $A = (1, 1)$. For the first step, perform the transformation $T(\lambda_0^{(1)}, \lambda_1^{(1)})$ on the line OA. Let the broken line consisting of two straight line segments represent the function F_1. On each of these two straight line segments perform the transformation $T(\lambda_0^{(2)}, \lambda_1^{(2)})$ to obtain 2^2 straight line segments representing a function F_2. On these straight line segments we perform the transformation $T(\lambda_0^{(3)}, \lambda_1^{(3)})$ to get F_3, and so on. Let

$$\lambda_0^{(k)} = \frac{1 - r_k}{2},$$

$$\lambda_1^{(k)} = \frac{1 + r_k}{2},$$

where $-1 < r_k < 1$ for each k. Let

$$\mu_p = \prod_{k=1}^{p}\left(\frac{1 + |r_k|}{2}\right).$$

Assume that the series

$$\sum_{p=1}^{\infty} \mu_p$$

converges. Then F_p converges uniformly to a continuous function F; the argument is the same as for the previous example. Similarly, the same reasoning shows that if

$$x = \frac{\theta_1}{2} + \frac{\theta_2}{2} + \cdots + \frac{\theta_p}{2} + \cdots,$$

then

$$F(x) = \theta_1 \lambda_0^{(1)} + \theta_2 \lambda_{\theta_1}^{(1)}\lambda_0^{(2)} + \theta_3 \lambda_{\theta_1}^{(1)}\lambda_{\theta_2}^{(2)}\lambda_0^{(3)} + \cdots.$$

Also, if x and $x' > x$ have the same first p digits in their dyadic representations, then

$$F(x') - F(x) < \lambda_{\theta_1}^{(1)}\lambda_{\theta_2}^{(2)} \cdots \lambda_{\theta_p}^{(p)} = \frac{1}{2^p} \prod_{k=1}^{p} (1 - \varepsilon_k r_k),$$

where

$$\varepsilon_k = \begin{cases} 1 & \text{if } \theta_k = 0 \\ -1 & \text{if } \theta_k = 1. \end{cases}$$

The interested reader may consult [S2] for the proof of the following theorem.

Theorem 8.6.6. *The function F is purely singular if and only if the series*

$$\sum_{k=1}^{\infty} r_k^2$$

diverges.

Another method of constructing a strictly increasing singular function is provided by the following example.

EXAMPLE 8.6.7 [T1]. We define a function F on $[0, 1]$ such that $F(0) = 0$. If $x \in (0, 1]$, write the dyadic representation of x as

(8.6.8)
$$x = \sum_{r=0}^{\infty} 2^{-a_r},$$

where a_r are positive integers ordered such that $a_0 < a_1 < a_2 < \cdots < a_r < \cdots$. For example, if $x = 1$, then $a_r = r$. Let ρ be a fixed positive number. Define

(8.6.9)
$$F(x) = \sum_{r=0}^{\infty} \rho^r (1 + \rho)^{-a_r},$$

for $x \in (0, 1]$ given by (8.6.8). Clearly, $F(1) = 1$. We shall show that if $\rho \neq 1$, then F is strictly increasing, continuous, and its derivative is zero almost everywhere in the interval $[0, 1]$.

Let $0 < \rho < \infty$ and $0 < x \leq 1$. By (8.6.9), we have $F(0) < F(x)$. Now let $0 < x < y \leq 1$, where x is given by (8.6.8) and

(8.6.10)
$$y = \sum_{r=0}^{\infty} 2^{-b_r},$$

where $b_0 < b_1 < \cdots < b_r \cdots$ are positive integers. Let s be the smallest integer such that $a_s \neq b_s$. Since $b_s < a_s$, we have

$$\sum_{r=s}^{\infty} \rho^r (1 + \rho)^{-a_r} \leq \rho^s (1 + \rho)^{-a_s+1} \leq \rho^s (1 + \rho)^{-b_s} < \sum_{r=s}^{\infty} \rho^r (1 + \rho)^{-b_r},$$

which implies that $F(x) < F(y)$; thus, F is strictly increasing.

If $x \in (0, 1]$ is given by (8.6.8), define

$$(8.6.11) \qquad x_n = \sum_{a_r \leq n} 2^{-a_r} \quad \text{and} \quad y_n = x_n + 2^{-n}$$

for $n \geq 0$. Then $x_n < x \leq y_n$ and

$$(8.6.12) \qquad F(y_n) - F(x_n) = \rho^{k_n}(1 + \rho)^{-n},$$

where k_n is the number of subscripts $r = 0, 1, 2, \ldots, n$ for which $a_r \leq n$. In (8.6.12), $0 \leq k_n \leq n + 1$; therefore,

$$\lim_{n \to \infty} [F(y_n) - F(x_n)] = 0.$$

This implies that F is continuous on $(0, 1]$. Since

$$\lim_{x \to 0} F(x) = F(0),$$

we have continuity on $[0, 1]$.

If $\rho = 1$, then $F(x) = x$ for $x \in [0, 1]$. We shall show that if F has a finite nonzero derivative for some $x \in (0, 1]$, then ρ must be 1. If $x \in (0, 1]$ and $F'(x)$ exists, by (8.6.11) and (8.6.12), we have

$$F'(x) = \lim_{n \to \infty} \frac{F(y_n) - F(x_n)}{y_n - x_n} = \lim_{n \to \infty} \left(\frac{2}{1 + \rho}\right)^n \rho^{k_n}.$$

If $F'(x)$ is finite and nonzero, it follows that

$$\lim_{n \to \infty} \rho^{k_n - k_{n-1}} = \frac{1 + \rho}{2}.$$

Thus, either $\rho = 1$ or

$$\lim_{n \to \infty} (k_n - k_{n-1}) = k.$$

If the latter case is true, then

$$\lim_{n \to \infty} \frac{k_n}{n} = k$$

and $0 \leq k_n \leq n + 1$, which implies $k = 0$ or $k = 1$. This, together with $\rho^k = (1 + \rho)/2$, implies again that $\rho = 1$.

Hence, if $\rho \neq 1$, then $F' = 0$ wherever it exists finitely. But since F is increasing, we know that F' exists finitely a.e.

In the generalization of Example 8.6.1, if we choose $r_k = (1 - \rho)/(1 + \rho)$ where $0 < \rho < 1$, we obtain the function given by (8.6.9).

EXAMPLE 8.6.13 [K3]. Let $f: [a, b] \to [c, d]$ be increasing and onto. If there exists a set $E \subset [a, b]$ such that $mE = 0$ and $m[f(E)] = d - c$, then f is singular. Clearly, f is continuous and differentiable a.e. Without loss of generality, suppose $a, b \in E$. If $f'(x) \geq M > 0$ on a set with measure $r > 0$, then $f'(x) \geq M$ on a closed set $B \supset [a, b] - E$ with $mB \geq r/2$. For $x \in B$ there is a

$\delta_x > 0$ such that

$$\frac{f(w) - f(x)}{w - x} \geq \frac{M}{2}$$

if $|w - x| < \delta_x$. Let $\varepsilon > 0$ be given and $f(E)$ and $f(B)$ be disjoint. There exists a family $I(t)$ of open intervals such that

$$\sum_t m[I(t)] < \varepsilon$$

and

$$f(B) \subset \bigcup_t I(t).$$

If $x \in B$, there exists t_x such that $f(x) \in I(t_x)$. Since $f'(x) \geq M$ and f is continuous at x, we can pick $a_x, b_x \in [a, b]$ such that

$$f(x) \in (f(a_x), f(b_x)) \subset I(t_x)$$

and $a_x < x < b_x$ with $x - a_x < \delta_x$ and $b_x - x < \delta_x$. Then

$$f(b_x) - f(a_x) \geq M \frac{b_x - a_x}{2}.$$

For each $x \in B$, pick (a_x, b_x) in this manner to form a covering of B. Because B is closed, we can take a minimal finite subcovering $J_1, J_2, \ldots, J_n$. Since this is a minimal subcovering, no point in B is in more than two of these intervals, so

$$2\varepsilon > 2 \sum_t m[I(t)] \geq M \sum_{i=1}^n mJ_i \geq M \cdot mB \geq \frac{Mr}{2}.$$

Thus, $f' = 0$ a.e. on $[a, b]$.

EXERCISES

In Exercises 1–11, α, β, $P_{\alpha\beta}$, and $\omega_{\alpha\beta}$ are as defined in §8.3.

1. Show that $P_{\alpha\beta}^c$ has measure one.

2. Show that $P_{\alpha\beta}$ is perfect, nowhere dense, and has measure zero.

3. Show that $P_{\alpha\beta}^c$ is open and consists of all the open nonoverlapping intervals (ξ', ξ'') with the endpoints

$$\xi' = \frac{a_1'}{\alpha} + \cdots + \frac{a_{k-1}'}{\alpha^{k-1}} + \frac{a_k' - q + 1}{\alpha^k},$$

$$\xi'' = \frac{a_1'}{\alpha} + \cdots + \frac{a_{k-1}'}{\alpha^{k-1}} + \frac{a_k'}{\alpha^k},$$

where the a_i are integral multiples of q and $a_k' \neq 0$.

4. Show $\omega_{\alpha\beta}(x)$ is single-valued for $x \in [0,1]$ even if x admits more than one representation.

5. Show $\omega_{\alpha\beta}$ is increasing, $\omega_{\alpha\beta}(0) = 0$, and $\omega_{\alpha\beta}(1) = 1$.

6. Show that $\omega_{\alpha\beta}$ has the intervals (ζ', ζ'') as defined in Exercise 3 as intervals of constancy.

7. Show $\omega_{\alpha\beta}(x) + \omega_{\alpha\beta}(1 - x) = 1$.

8. Show $\omega_{\alpha\beta}$ is continuous.

9. Show $\omega_{\alpha\beta}$ is not absolutely continuous.

10. Show that $\omega_{\alpha\beta}$ satisfies a Lipschitz condition of order

$$\mu = \frac{\log \beta}{\log \alpha}, \quad \beta = \alpha^\mu,$$

the Lipschitz coefficient being not greater than

$$c = \beta(q - 1)^{-\mu}.$$

Show that these constants are the best possible.

11. Divide $P_{\alpha\beta}$ into three subsets, $P_{\alpha\beta}^0$, $P_{\alpha\beta}^+$, and $P_{\alpha\beta}^-$ where

$P_{\alpha\beta}^0$ consists of all $x \in P_{\alpha\beta}$ having an infinite number of digits different from 0 and from $\alpha - 1$;

$P_{\alpha\beta}^+$ consists of all $x \in P_{\alpha\beta}$ with only a finite number of digits different from 0;

$P_{\alpha\beta}^-$ consists of all $x \in P_{\alpha\beta}$ with only a finite number of digits different from $\alpha - 1$.

(a) Show $\omega'_{\alpha\beta}(x) = 0$ at all points $x \in P_{\alpha\beta}^c$.
(b) Show $D^+\omega_{\alpha\beta}(x) = \infty$ and $D^-\omega_{\alpha\beta}(x) = 0$ at all points $x \in P_{\alpha\beta}^+$.
(c) Show $D^+\omega_{\alpha\beta}(x) = 0$ and $D^-\omega_{\alpha\beta}(x) = \infty$ at all points $x \in P_{\alpha\beta}^-$.
(d) Show that if $x \in P_{\alpha\beta}^0$, one of the following cases arises:
    (i) There is a unique derivative (∞).
    (ii) The derivative on the right is ∞, whereas there is no derivative on the left.
    (iii) The derivative on the left is ∞, whereas there is no derivative on the right.
    (iv) There is no derivative on the right or left.

Exercises 12–19 refer to §8.4.

12. Prove that the set E is closed, perfect, uncountable, nowhere dense, and has Lebesgue measure zero.

13. Show that ϕ is constant on every closed D_n-interval and that the union of the intervals of constancy of ϕ is a set of measure one.

14. Show that ϕ is an increasing function of x in $[0,1]$; moreover, when $0 \leq x < x' \leq 1$, then $\phi(x) \leq \phi(x')$, equality holding if and only if x and x' lie in the same closed D_n-interval.

15. Show that the function $y = \phi(x)$ assumes every value of y in the interval $[0,1]$. If $y \in (0,1]$ is rational, show that y is assumed by ϕ in one whole D_n-interval, whereas $y = 0$ and every irrational $y \in (0,1)$ are assumed at one point x.

16. Show that ϕ is continuous in $[0, 1]$.

17. For every positive λ, the λ-variation of ϕ in $[0, 1]$ is 1; hence, ϕ is not absolutely continuous. [**Hint**: First show that if a function f is monotone in the closed interval $[a, b]$ and if the intervals of constancy of f cover $[a, b]$ except for a set of measure zero, then the λ-variation of f in $[a, b]$ equals $|f(b) - f(a)|$ for every positive λ. The result will then follow.]

18. Show that the curve $y = \phi(x)$ is of arc length 2. [**Hint**: First show that if a function f is monotone in the closed interval $[a, b]$ and if its λ-variation in $[a, b]$ equals $|f(b) - f(a)|$, then the arc length of the curve $y = f(x)$ is $b - a + |f(b) - f(a)|$.]

19. Show that if $x \in D$, then $\phi'(x) = 0$. (For other derivatives of the function ϕ, see [HB].)

20. Verify that the Cantor-like set P of §8.5 is closed, perfect, and nowhere dense.

21. Let C be the Cantor ternary set. Let $I \subset R$ be an interval having finite positive measure. Prove that there exists an open interval $J \subset I \cap C^c$ such that

$$\frac{mJ}{mI} \geq \frac{1}{5}.$$

22. Let r be a dyadic rational in $(0, 1)$ and suppose that the associated open interval deleted in the construction of the Cantor-like set of Example 8.5.4 is $I_r = (a, b)$. Show that corresponding to the dyadic rational, $1 - r$ is the interval $I_{1-r} = (1 - b, 1 - a)$.

23. Suppose f is an increasing function defined on $[0, 1]$ with $f(0) = 0$ and $f(1) = 1$. Let L be the arc length of the curve $(x, f(x))$. Show that $L \leq 2$ with equality if and only if $f'(x) = 0$ a.e.

CHAPTER 9

Spaces of *BV* and *AC* Functions

Functions of bounded variation and functions which are absolutely continuous play an important role in applications such as partial differential equations and calculus of variations. To study the properties of these classes of functions requires the introduction of a norm on each space. One such norm on *BV* was introduced in §6.3. This chapter is devoted to discussing other norms on these two function spaces. The topologies induced by these norms have ramifications in the study of the concepts of bounded variation and absolute continuity for functions of several variables. We begin first with some convergence results.

§9.1. Convergence in Variation

The results of this section are from [AC].

Definition 9.1.1. Let $\{f_n\}$ be a sequence of functions defined on the interval $[a, b]$. Suppose f_n converges pointwise to a function f_0 which is of bounded variation and the total variation $T_{f_n}[u, b]$ converges to $T_{f_0}[a, b]$. Then we will say that f_n *converges in variation* to f_0 and the notation $f_n - v \to f_0$ will be used.

Theorem 9.1.2. *If* $f_n(a) \to f_0(a)$, $f_n(b) \to f_0(b)$, *and* $T_{f_n}[a, b] \to T_{f_0}[a, b]$, *then* $P_n \to P_0$ *and* $N_n \to N_0$ *where* P_n *and* N_n *denote the total positive and negative variations, respectively, of* f_n *on* $[a, b]$.

216

Proof. Since

$$f_n(b) = f_n(a) + P_n - N_n,$$

this result follows.   □

Theorem 9.1.3. *If $f_n \to f_0$ pointwise on $[a, b]$, then*

$$T_{f_0}[a, b] \le \varlimsup_{n \to \infty} T_{f_n}[a, b].$$

Corollary 9.1.4. *If $f_n - v \to f_0$ on $[a, b]$, then $f_n - v \to f_0$ on every subinterval of $[a, b]$.*

Theorem 9.1.5. *If $f_n \to f_0$ on a set of points which is everywhere dense in $[a, b]$, $T_{f_n}[a, b] \to T_{f_0}[a, b]$, and f_0 is continuous on $[a, b]$, then $f_n \to f_0$ everywhere on $[a, b]$.*

Theorem 9.1.6. *If $f_n - v \to f_0$ on $[a, b]$ and $f_0(x') = f_0(x'-)$ [or $f_0(x') = f_0(x'+)$], then x' is a point of uniform convergence on the left (or right) for both f_n and $T_{f_n}[a, x]$.*

Proof. Since $T_{f_0}[a, x]$ is continuous on the left for any $x' \in (a, b]$, then for any $\varepsilon > 0$, we can find $\delta > 0$ such that $T_{f_0}[x' - \delta, x'] < \varepsilon$. Then we can choose N so that for $n > N$, we have

$$|f_n(x') - f_0(x')| < \varepsilon,$$

$$|T_{f_n}[a, x' - \delta] - T_{f_0}[a, x' - \delta]| < \varepsilon,$$

$$T_{f_n}[x' - \delta, x'] < 2\varepsilon.$$

Thus, $|f_n(x) - f_0(x)| < 4\varepsilon$ and $|T_{f_n}[a, x] - T_{f_0}[a, x]| < 4\varepsilon$ for $0 \le x' - x \le \delta$, $n > N$.   □

Corollary 9.1.7. *The hypotheses of the previous theorem imply that the convergence of f_n to f_0 and $T_{f_n}[a, x]$ to $T_{f_0}[a, x]$ is uniform on $[a, b]$.*

Theorem 9.1.8. *If $f_n - v \to f_0$ on $[a, b]$ and $f_0(x') \ne f_0(x'-)$ or $[f_0(x') \ne f_0(x'+)]$, then x' is a point of uniform convergence on the left (or right) for both f_n and $T_{f_n}[a, x]$ or for neither, depending on whether $f_n(x'-)$ [or $f_n(x'+)$] tends to $f_0(x'-)$ [or $f_0(x'+)$] or not.*

Proof. Let $\bar{f}_n(x) = f_n(x)$ for $a \le x < x'$, $\bar{f}_n(x') = f_n(x'-)$, for $n = 0, 1, 2, \ldots$. Then

$$T_{f_n}[a, x] = T_{\bar{f}_n}[a, x] + |f_n(x) - \bar{f}_n(x)|$$

for each n and $a \le x < x'$. If x' is a point of uniform convergence on the left for either f_n or $T_{f_n}[a, x]$, then $\bar{f}_n(x') \to \bar{f}_0(x')$. By Theorem 9.1.6, x' is a point of uniform convergence on the left for both $\bar{f}_n$ and $T_{\bar{f}_n}[a, x]$, and hence for both f_n and $T_{f_n}[a, x]$. $\square$

Corollary 9.1.9. *If $f_n - v \to f_0$ on $[a, b]$, a necessary and sufficient condition that $T_{f_n}[a, x] \to T_{f_0}[a, x]$ uniformly on $[a, b]$ is that $f_n \to f_0$ uniformly on $[a, b]$.*

Remark. If $T_{f_n}[a, x]$ converges uniformly, the same is true of the total positive and negative variations, $P_{f_n}[a, x]$ and $N_{f_n}[a, x]$.

Lemma 9.1.10. *Let $P = \{a = x_0, x_1, \ldots, x_p = b\}$ be any partition of $[a, b]$. Let P' be a new partition obtained by adding a single point x' to P. Let*

$$\sum (f, P) = \sum_{k=1}^{p} |\Delta f_k|.$$

If f is finite and $f(x) > \alpha > 0$ [or $f(x) < -\alpha < 0$] for $x \in [a, b]$, then

$$0 \le \sum \left(\frac{1}{f}, P'\right) - \sum \left(\frac{1}{f}, P\right) \le \frac{\sum (f, P') - \sum (f, P)}{\alpha^2}.$$

Theorem 9.1.11. *If $f_n - v \to f_0$ on $[a, b]$, $|f(x)| > 2\alpha > 0$ for $x \in [a, b]$, and f does not change sign in the interval $[a, b]$, then $1/f_n - v \to 1/f_0$.*

Proof. It can be shown that for n sufficiently large, then functions f_n are uniformly $> \alpha$ (or $< -\alpha$). Thus, without loss of generality, we may assume that f_n is uniformly $> \alpha$ and of uniformly bounded variation. Then there exists a double sequence of partitions P_n^p ($n, p = 0, 1, 2, \ldots$) such that for each p we have

$$P_n^p \subset P_{n+1}^p, \qquad P_n^p \subset P_n^{p+1} \qquad (n = 0, 1, 2, \ldots);$$

$$\sum (f_p, P_n^p) \to T_{f_p}[a, b], \qquad \sum \left(\frac{1}{f_p}, P_n^p\right) \to T_{(1/f_p)}[a, b]$$

as $n \to \infty$. Let $P_n = P_n^n$,

$$a_{mn} = \sum (f_m, P_n), \qquad b_{mn} = \sum \left(\frac{1}{f_m}, P_n\right).$$

Because of the choice of partitions and the fact that $f_n - v \to f_0$, a_{mn} is non-decreasing in n for each m and the iterated limits as $m, n \to \infty$ both exist and are equal. By a lemma of Hildebrandt [H1], $\lim_{n \to \infty} a_{mn}$ exists uniformly in m. From the previous lemma, we infer that $\lim_{n \to \infty} b_{mn}$ exists uniformly in m. By our choice of partitions, we have

$$\lim_{m \to \infty} b_{mn} = \sum \left(\frac{1}{f}, P_n\right).$$

Thus,

$$\lim_{n \to \infty} \lim_{m \to \infty} b_{mn} = T_{1/f}[a, b] = \lim_{m \to \infty} \lim_{n \to \infty} b_{mn} = \lim_{m \to \infty} T_{(1/f_m)}[a, b]. \qquad \square$$

Remark. The requirement that f be of fixed sign in $[a, b]$ cannot be deleted from the hypotheses of the previous theorem. For if f is allowed to change signs in $[a, b]$, the fact that $|f|$ is bounded away from zero does not imply that f_n is uniformly bounded away from zero, for sufficiently large n. But even the condition that f_n be so bounded is not sufficient for the result of the theorem to hold, as illustrated by the following example.

EXAMPLE 9.1.12. Let

$$f_0(x) = 1 \quad \text{for } 0 \le x < 1, f_0(1) = -1.$$

For $n = 1, 2, 3, \ldots$, let f_n be a monotone function which decreases from 1 to -1 and continuous except for a jump from $\frac{1}{2}$ to $-\frac{1}{2}$ somewhere in the interval such that $f_n \to f_0$.

Remark. If $f_n - v \to f_0$ and $g_n - v \to g_0$ on $[a, b]$, we do not necessarily have $f_n + g_n - v \to f_0 + g_0$ or $f_n g_n - v \to f_0 g_0$. In fact, convergence of the sum and product do not necessarily follow with the stronger assumption that f_0 and g_0 are continuous on $[a, b]$ (which by Corollary 9.1.7 implies uniformity of convergence). Even with the stronger assumption that $f_n - v \to f_0$ and g_0 is a function of bounded variation, we cannot conclude that $f_n + g_0 - v \to f_0 + g_0$ or $f_n g_0 - v \to f_0 g_0$.

EXAMPLE 9.1.13. Let $g_0(x) = -x$ for $x \in [0, 1]$; for $n = 1, 2, 3, \ldots$, let

$$f_n\left(\frac{m}{2^n}\right) = \begin{cases} \dfrac{m}{2^n}, & m = 0 \bmod 2 \\[2mm] \dfrac{m}{2^n} + \dfrac{1}{2^{n+1}}, & m = 1 \bmod 4 \\[2mm] \dfrac{m}{2^n} - \dfrac{1}{2^{n+1}}, & m = 3 \bmod 4, \end{cases}$$

and between the points $m/2^n$, let f_n be defined linearly. Note that f_n is a sequence of absolutely continuous functions which converge in variation to an absolutely continuous function f_0 and g_0 is absolutely continuous; however, $f_n + g_0$ does not converge in variation to $f_0 + g_0$.

Theorem 9.1.14. *Let f_n be a sequence of absolutely continuous functions converging to a limit function f_0 on $[a, b]$. Let f_n' converge asymptotically to a limit function and let f_n' $(n = 1, 2, 3, \ldots)$ be dominated by a summable function. Then $f_n - v \to f_0$ on $[a, b]$.*

Proof. Because each $f_n = 1, 2, 3, \ldots$, is absolutely continuous,

$$T_{f_n}[a, b] = \int_a^b |f_n'(x)| \, dx.$$

By the hypotheses of the theorem, f_0 is also absolutely continuous and we may pass to the limit under the integral sign.   □

Corollary 9.1.15. *Let the series*

$$\sum_{i=0}^{\infty} a_i x^i,$$

with real coefficients, have the radius of convergence $R > 0$; let the sum of the series be denoted by $S(x)$ and let

$$S_n(x) = \sum_{i=0}^{n} a_i x^i.$$

Then $S_n - v \to S$ on each interval $[a, b]$ for $-R < a < b < R$.

§9.2. Convergence in Length

The results of this section are from [AL].

Definition 9.2.1. Let $f: [a, b] \to \mathbf{R}^1$. Let $P = \{a = x_0, x_2, \ldots, x_p = b\}$ be a partition of $[a, b]$. We will denote by B a broken line inscribed in the curve $y = f(x)$ and consisting of p segments, the ith segment ($i = 1, 2, \ldots, p$) having endpoints at $(x_{i-1}, f(x_{i-1}))$ and $(x_i, f(x_i))$ and length b_i. The function whose graph is B will be denoted by $B(x)$. Broken lines inscribed in two distinct curves and determined by the same partition are called *corresponding*. A curve (continuous or not) is said to have *finite length in the sense of Peano* if the lengths of all inscribed broken lines B are bounded; then their least upper bound is the length of the curve and will be denoted by

$$L_f[a, b].$$

Definition 9.2.2. Let f_n ($n = 1, 2, 3, \ldots$) be a sequence of functions defined on an interval $[a, b]$. Suppose f_n converges pointwise to a limit function f_0 which is of bounded variation. If $L_{f_n}[a, b] \to L_f[a, b]$, then we say that f_n *converges in length* to f_0 and we use the notation $f_n - L \to f_0$.

Theorem 9.2.3. *If $f_n \to f_0$ on $[a, b]$, then*

$$L_{f_0}[a, b] \leq \varliminf_{n \to \infty} L_{f_n}[a, b].$$

Corollary 9.2.4. *If $f_n - L \to f_0$ on $[a, b]$, then $f_n - L \to f_0$ on every subinterval of $[a, b]$.*

Theorem 9.2.5. *If $f_n \to f_0$ on a set of points everywhere dense in $[a, b]$, $L_{f_n}[a, b] \to L_{f_0}[a, b]$, and f_0 is continuous, then $f_n \to f_0$ everywhere on $[a, b]$.*

Theorem 9.2.6. *If $f_n - L \to c$ on $[a,b]$, where c is a constant, then $f_n - v \to c$ on $[a,b]$. The converse is also true.*

Proof. If f is of bounded variation on $[a,b]$, then

$$(9.2.7) \quad b - a + T_f[a,b] \geq L_f[a,b] \geq \{(b-a)^2 + (T_f[a,b])^2\}^{1/2}.$$

The result then follows. □

Corollary 9.2.8. *If f is of bounded variation on $[a,b]$, then $L_f[a,x]$ is continuous on the right (or left) at each x where f is continuous on the right (or left). If f is absolutely continuous on $[a,b]$, both $T_f[a,x]$ and $L_f[a,x]$ are also.*

Theorem 9.2.9. *If $f \in BV[a,b]$ (continuous or not), there is always a sequence of inscribed broken lines B_n ($n = 1,2,3,\ldots$) such that $B_n - L \to f$ and $B_n - v \to f$.*

Proof. We can choose the sequence of partitions P_n determining B_n so that $P_n \subset P_{n+1}$ for each n and

$$P = \bigcup_{n=1}^{\infty} P_n$$

is everywhere dense in $[a,b]$ and contains all the points of discontinuity of f. Then $B_n \to f$ pointwise on $[a,b]$: for $\xi' \in P$, the convergence is evident, whereas if $\xi \notin P$, we have

$$\varlimsup_{n\to\infty} |f(\xi) - B_n(\xi)| \leq T_f[\xi',\xi] + \varlimsup_{n\to\infty} |f(\xi') - B_n(\xi')| + \varlimsup_{n\to\infty} T_{B_n}[\xi',\xi]$$

$$\leq 2T_f[\xi',\xi'']$$

for $\xi' < \xi < \xi''$ and ξ', $\xi'' \in P$. From Theorem 9.2.3 and the definitions of L and T, it follows that $L_{B_n}[a,b] \to L_f[a,b]$ and $T_{B_n}[a,b] \to T_f[a,b]$. □

Theorem 9.2.10. *If $f_n - L \to f_0$ on $[a,b]$, then $f_n - v \to f_0$; the converse is not true even when the sequence f_n is absolutely continuous.*

Proof. Let $\varepsilon > 0$ and let B be a broken line inscribed in f_0 determined by the partition $P = \{a = x_0, x_1, \ldots, x_p = b\}$ such that

$$L_B[a,b] > L_{f_0}[a,b] - \varepsilon.$$

Let

$$\varepsilon_i = L_{f_0}[x_{i-1}, x_i] - b_i,$$

$$d_i = x_i - x_{i-1}.$$

Then clearly $\sum \varepsilon_i < \varepsilon$ and there exists an m such that

$$L_{f_n}[x_{i-1}, x_i] \leq L_{f_0}[x_{i-1}, x_i] + \varepsilon_i \quad (i = 1, 2, \ldots, p; n > m).$$

From (9.2.7), we have for each i

$$T_{f_n}[x_{i-1}, x_i] \leq \{(L_{f_n}[x_{i-1}, x_i])^2 - d_i^2\}^{1/2}$$
$$\leq [(b_i + 2\varepsilon_i)^2 - d_i^2]^{1/2}$$
$$\leq (b_i^2 - d_i^2)^{1/2} + 2(b_i\varepsilon_i)^{1/2} + 2\varepsilon_i.$$

Then, by Schwarz's inequality, we have

$$T_{f_n}[a, b] \leq T_{f_0}[a, b] + 2\left(\sum_{i=1}^{p} b_i \sum_{i=1}^{p} \varepsilon_i\right)^{1/2} + 2\sum_{i=1}^{p} \varepsilon_i$$
$$\leq T_{f_0}[a, b] + 2(\varepsilon L_{f_0}[a, b])^{1/2} + 2\varepsilon.$$

Applying Theorem 9.2.3, we have the desired result. Example 9.1.13 illustrates that the converse of the theorem does not hold, even when all the functions involved are absolutely continuous.   □

Corollary 9.2.11. $L_f[a, b] = L_{T_f[a, x]}[a, b]$.

Proof. From (9.2.7), it follows that both these quantities are finite or neither is finite. In the first case, the inequality

$$|f(x_i) - f(x_{i-1})| \leq T_f[a, x_i] - T_f[a, x_{i-1}]$$

demonstrates that, of two corresponding broken lines inscribed in f and $T_f[a, x]$, the former has length no greater than the latter. Hence,

$$L_f[a, b] \leq L_{T_f[a, x]}[a, b].$$

On the other hand, for any broken line B, we have

$$L_B[a, b] = L_{T_B[a, x]}[a, b];$$

hence, choosing a sequence of broken lines B_n ($n = 1, 2, 3, \ldots$) with $B_n - L \to f$, we get

$$L_f[a, b] = \lim_{n \to \infty} L_{B_n}[a, b] = \lim_{n \to \infty} L_{T_{B_n}[a, x]}[a, b] \geq L_{T_f[a, x]}[a, b],$$

since $T_{B_n}[a, x] \to T_f[a, x]$ by the previous theorem.   □

Using Theorem 9.2.10 and the results of §9.1, the reader may easily prove the next four results.

Theorem 9.2.12. If $f_n - L \to f_0$ on $[a, b]$ and $f_0(x') = f_0(x'-)$ [or $f_0(x') = f_0(x'+)$], then x' is a point of uniform convergence on the left (or right) for f_n, $T_{f_n}[a, x]$, and $L_{f_n}[a, x]$.

Corollary 9.2.13. If $f_n - L \to f_0$ on $[a, b]$ and f_0 is continuous, then $f_n \to f_0$ uniformly on $[a, b]$, $T_{f_n}[a, x] \to T_{f_0}[a, x]$ uniformly on $[a, b]$, and $L_{f_n}[a, x] \to L_{f_0}[a, x]$ uniformly on $[a, b]$.

Theorem 9.2.14. *If $f_n - L \to f_0$ on $[a,b]$ and $f_0(x') \neq f_0(x'-)$ [or $f_0(x') \neq f_0(x'+)$], then x' is a point of uniform convergence on the left (or right) for f_n, $T_{f_n}[a,x]$, and $L_{f_n}[a,x]$, or for none of the three, depending on whether $f_n(x'-)$ [or $f_n(x'+)$] tends to $f_0(x'-)$ [or $f_0(x'+)$] or not.*

Corollary 9.2.15. *If $f_n - L \to f_0$ on $[a,b]$, a necessary and sufficient condition that $L_{f_n}[a,x] \to L_{f_0}[a,x]$ uniformly on $[a,b]$ is that $f_n \to f_0$ uniformly on $[a,b]$.*

Theorem 9.2.16. *If $f, g \in BV[a,b]$, then*

(9.2.17) $$T_{f+g}[a,b] \leq T_f[a,b] + T_g[a,b];$$

(9.2.18) $$L_{f+g}[a,b] \leq L_f[a,b] + L_g[a,b];$$

(9.2.19) $$T_{f \cdot g}[a,b] \leq M_2 T_f[a,b] + M_1 T_g[a,b],$$

where $M_1 = \sup\{|f(x)|: x \in [a,b]\}$ and $M_2 = \sup\{|g(x)|: x \in [a,b]\}$,

(9.2.20) $$\frac{(T_{f-mx}[a,b])^2}{1+m^2} \leq (L_f[a,b])^2 - (b-a)^2(1+m^2),$$

where

$$m = \frac{f(b) - f(a)}{b - a}.$$

Proof. Inequalities (9.2.17) and (9.2.19) follow easily from the definition of bounded variation. Inequality (9.2.18) follows from the fact that the ith segment of a broken line inscribed in $f + g$ has length no greater than the corresponding segment inscribed in f plus the quantity $|g(x_i) - g(x_{i-1})|$. Now we consider the proof of (9.2.20) for the case of a broken line B, denoting the derivative of $B(x)$ by $\phi(x)$ so that

$$B(x) = \int_a^x \phi(t)\, dt.$$

Then (9.2.20) becomes

(9.2.21) $$\left[\int_a^b (1 + \phi^2)^{1/2}\, dx\right]^2 \geq (b-a)^2(1 + m^2) + \frac{[\int_a^b |\phi - m|\, dx]^2}{1 + m^2},$$

where

$$m = \frac{\int_a^b \phi\, dx}{b - a}.$$

Designate the integral over the set of points x for which $\phi - m \geq 0$ by $\int_+$; designate the integral over the set of points x for which $\phi - m < 0$ by $\int_-$. By Minkowski's inequality, we have

$$\int_+ (1 + \phi^2)^{1/2}\, dx \geq \left[\left(\int_+ dx\right)^2 + \left(\int_+ \phi\, dx\right)^2\right]^{1/2},$$

together with a similar inequality for $\int_-$. Letting

$$\alpha = \int_+ dx,$$

$$\beta = \int_+ \phi\, dx,$$

$$\gamma = \int_- dx,$$

$$\delta = \int_- \phi\, dx,$$

we have $m = (\beta + \delta)/(\alpha + \gamma)$ and

$$\int_a^b |\phi - m|\, dx = \int_+ (\phi - m)\, dx - \int_- (\phi - m)\, dx$$

$$= \beta - m\alpha - \delta + m\gamma$$

$$= \frac{2(\beta\gamma - \alpha\delta)}{\alpha + \gamma},$$

and (9.2.21) will be true if

(9.2.22)

$$\sqrt{\alpha^2 + \beta^2} + \sqrt{\gamma^2 + \delta^2} \geq \left((\alpha + \gamma)^2 + (\beta + \delta)^2 + \frac{4(\beta\gamma - \alpha\delta)^2}{(\alpha + \gamma)^2 + (\beta + \delta)^2}\right)^{1/2}$$

holds. But under a suitable orthogonal transformation of the ξ, η-plane, the two points $\xi = \alpha$, $\eta = \beta$, and $\xi = \gamma$, $\eta = \delta$, respectively, go into (α', β') and (γ', δ') with $\beta' + \delta' = 0$, whereas (9.2.22) is invariant and reduces to the triangular inequality

$$\sqrt{\alpha'^2 + \beta'^2} + \sqrt{\gamma'^2 + \beta'^2} \geq \sqrt{(\alpha' + \gamma')^2 + (2\beta')^2}.$$

This proves (9.2.21) from which (9.2.20) follows at once by approximating f by a sequence of broken lines B_n with $B_n - L \to f$ and using Theorem 9.2.3. □

Remark. That convergence in variation is not invariant under the operations of addition and multiplication, even when all the functions involved are absolutely continuous, was shown in the previous section. Similarly, convergence in length is not invariant under addition, even when the limit function is continuous and all the approximating functions are absolutely continuous, as seen by the following example.

EXAMPLE 9.2.23. Let f_0 defined on $[0, 1]$ be the Cantor ternary function and let $f_0(x) = 1$ for $1 \leq x \leq 2$. If B is any broken line inscribed in f_0 on $[0, 1]$, we have

$$T_{f_0-B}[0,1] = \int_0^1 |B'|\,dx + T_{f_0}[0,1] = 2,$$

which shows that $f_n - L \to f_0$ on $[0,1]$, with f_0 continuous and f_n ($n = 1,2,3,\ldots$) absolutely continuous, does not imply that $T_{f_n-f_0}[0,1] \to 0$. In view of a theorem of Plessner [D2], there exists $\beta > 0$ and a sequence of positive numbers α_n ($n = 1,2,3,\ldots$) such that $\alpha_n \to 0$ as $n \to \infty$ and

$$T_{f_0(x)-f_0(x+\alpha_n)}[0,1] > \beta$$

for each n. Let a sequence of partitions P_n ($n = 1,2,3,\ldots$) of $[0,1]$ be chosen as follows: For each n the partition P'_n determines a sum which approximates $T_{f_0(x)-f_0(x-\alpha_n)}[0,1]$ within $\beta/2$; let the sequence of partitions P_n be chosen so that P_n includes $P_{n-1} \cup P'_n$ for each n and

$$P = \bigcup_{n=1}^{\infty} P_n$$

is everywhere dense in $[0,1]$. Designate by B_n, B_n^*, and C_n, respectively, the broken lines inscribed in $f_0(x)$, $f_0(x + \alpha_n)$, and $f_0(x) - f_0(x + \alpha_n)$ and determined by P_n. Then

$$B_n(x) - B_n^*(x) \equiv C_n(x),$$

$$B_n^*(x) - f_0(x) \equiv B_n(x) - C_n(x) - f_0(x),$$

$$B_n \to f_0,$$

$$|C_n(x)| \le \max|f_0(x) - f_0(x + \alpha_n)| \to 0,$$

whence $B_n^* \to f_0$. Also, by (9.2.3), we have

$$L_{f_0}[0,1] \le \varliminf_{n\to\infty} L_{B_n^*}[0,1] \le \varlimsup_{n\to\infty} L_{B_n^*}[0,1] \le \varlimsup_{n\to\infty} L_{f_0(x+\alpha_n)}[0,1] = L_{f_0}[0,1].$$

Hence, $B_n^* - L \to f_0$; also, $B_n - L \to f_0$. On the other hand,

$$T_{B_n-B_n^*}[0,1] = T_{C_n}[0,1] > \frac{\beta}{2}$$

for each n, which implies by Theorem 9.2.10 that $B_n - B_n^*$ does not converge in length to the zero function.

However, if the limit functions are absolutely continuous, the property of convergence in length is invariant under addition and multiplication independently of the nature of the approximating functions. To establish this, we need some preliminary results.

Lemma 9.2.24. *If f_0 is absolutely continuous on $[a,b]$, P an arbitrary partition of $[a,b]$, B_0 the broken line inscribed in f_0 and determined by P, and $\mathfrak{F}_h$ ($h = 1,2,\ldots,q$) those subintervals of P in which the slope m_h of B_0 is in absolute* *value* $\geq M$, *there exists a function* $\alpha(M)$ *independent of* P *such that*

$$\sum_h T_{f_0}[\mathfrak{F}_h] \leq \alpha(M)$$

and

$$\alpha(M) \to 0 \quad \text{as } M \to \infty.$$

Proof. First, observe that (letting $|\mathfrak{F}_h|$ denote the length of the subinterval $\mathfrak{F}_h$)

$$\sum |\mathfrak{F}_h| M \leq \sum |\mathfrak{F}_h| \cdot |m_h| \leq T_{f_0}[a,b];$$

hence,

$$\sum |\mathfrak{F}_h| \leq \frac{T_{f_0}[a,b]}{M}.$$

Then, because f_0 is absolutely continuous, $T_{f_0}[a,x]$ is absolutely continuous. So for any set of nonoverlapping intervals the sum of whose lengths is $\leq T_{f_0}[a,b]/M$, and in particular for the set of intervals $\mathfrak{F}_h$, we have

$$\sum_h T_{f_0}(\mathfrak{F}_h) \leq \alpha(M),$$

with $\alpha(M) \to 0$ as $M \to \infty$. $\qquad\square$

Lemma 9.2.25. *Under the hypotheses of the previous lemma, if* $f_n - v \to f_0$ *on* $[a,b]$, *then*

$$\varlimsup_{n\to\infty} \sum_h T_{f_n-f_0}(\mathfrak{F}_h) \leq \varlimsup_{n\to\infty} \sum_h T_{f_n}(\mathfrak{F}_h) + \sum_h T_{f_0}(\mathfrak{F}_h) \leq 2\alpha(M).$$

Proof. Since only a finite number of intervals $\mathfrak{F}_h$ are involved and $T_{f_n} \to T_{f_0}$ in each, the conclusion follows. $\qquad\square$

Theorem 9.2.26. *If* $f_n - L \to f_0$ *on* $[a,b]$ *and* f_0 *is absolutely continuous, then* $T_{f_n-f_0}[a,b] \to 0$.

Proof. Let P be any partition of $[a,b]$ determining a broken line B_n inscribed in f_n ($n = 0,1,2,\ldots$); for each subinterval determined by the partition, we have

$$\varlimsup_{n\to\infty} T_{f_n-f_0}[x_{i-1},x_i] \leq \varlimsup_{n\to\infty} T_{f_n-B_n}[x_{i-1},x_i] + \varlimsup_{n\to\infty} T_{B_n-B_0}[x_{i-1},x_i]$$

$$+ T_{f_0-B_0}[x_{i-1},x_i]$$

$$\leq \varlimsup_{n\to\infty} T_{f_n-B_n}[x_{i-1},x_i] + T_{f_0-B_0}[x_{i-1},x_i].$$

For those intervals $\mathfrak{F}_h$ for which the slope m_h of B_0 is numerically $\geq M$, we have by Theorem 9.2.10 and the previous lemma

$$(9.2.27) \qquad\qquad \varlimsup_{n\to\infty} \sum_h T_{f_n-f_0}[\mathfrak{F}_h] \leq 2\alpha(M),$$

where $\alpha(M) \to 0$ as $M \to \infty$. Let $\sum'$ denote the summation over those intervals in P for which the slope m_h of B_0 is numerically $< M$. We have by (9.2.20) and Schwarz's inequality

$$\sum' T_{f_0 - B_0}[x_{i-1}, x_i]$$

$$\le (1 + M^2)^{1/2} \sum' \{(L_{f_0}[x_{i-1}, x_i])^2 - (L_{B_0}[x_{i-1}, x_i])^2\}^{1/2}$$

(9.2.28) $\le (1 + M^2)^{1/2} \sum' (L_{f_0}[x_{i-1}, x_i] + L_{B_0}[x_{i-1}, x_i])^{1/2} \cdot$

$$(L_{f_0}[x_{i-1}, x_i] - L_{B_0}[x_{i-1}, x_i])^{1/2}$$

$$\le (1 + M^2)^{1/2} K(B_0),$$

where

$$K(B_0) = (L_{f_0}[a, b] + L_{B_0}[a, b])^{1/2} (L_{f_0}[a, b] - L_{B_0}[a, b])^{1/2}.$$

Since $B_n \to B_0$, we can sum over these same intervals to get

$$\overline{\lim_{n \to \infty}} \sum' T_{f_n - B_n}[x_{i-1}, x_i]$$

$$\le (1 + M^2)^{1/2} \overline{\lim_{n \to \infty}} (L_{f_n}[a, b] + L_{B_n}[a, b])^{1/2} (L_{f_n}[a, b] - L_{B_n}[a, b])^{1/2}.$$

Because $L_{f_n}[a, b] \to L_{f_0}[a, b]$ and $L_{B_n}[a, b] \to L_{B_0}[a, b]$, we obtain

(9.2.29) $\overline{\lim_{n \to \infty}} \sum' T_{f_n - B_n}[x_{i-1}, x_i] \le (1 + M^2)^{1/2} K(B_0).$

From (9.2.27), (9.2.28), and (9.2.29), we get

$$\overline{\lim_{n \to \infty}} T_{f_n - f_0}[a, b] \le 2(1 + M^2)^{1/2} K(B_0) + 2\alpha(M).$$

This inequality is independent of the partition P. $\alpha(M)$ can be made arbitrarily small by choosing M large enough. Once M has been chosen, the first term on the right can be made arbitrarily small by a suitable choice of a partition. Hence,

$$\overline{\lim_{n \to \infty}} T_{f_n - f_0}[a, b] = 0,$$

and the theorem is proved.   □

Theorem 9.2.30. *If $f_n \to f_0$ on $[a, b]$, $T_{f_n - f_0}[a, b] \to 0$, and $f_0 \in BV[a, b]$, then $f_n - L \to f_0$.*

Proof. By Theorem 9.2.3 and (9.2.18), we have

$$L_{f_0}[a, b] \le \underline{\lim_{n \to \infty}} L_{f_n}[a, b] \le \overline{\lim_{n \to \infty}} L_{f_n}[a, b]$$

$$\le L_{f_0}[a, b] + \overline{\lim_{n \to \infty}} T_{f_n - f_0}[a, b] = L_{f_0}[a, b].$$

Theorem 9.2.31. *If* $f_n - L \rightarrow f_0$ *and* $g_n - L \rightarrow g_0$ *on* $[a, b]$ *and* f_0, $g_0 \in$ *AC*$[a, b]$, *then* $[f_n + g_n] - L \rightarrow [f_0 + g_0]$ *and* $f_n g_n - L \rightarrow f_0 g_0$.

Proof. From (9.2.17), (9.2.19), and Theorem 9.2.26, we have

$$T_{f_n + g_n - f_0 - g_0}[a, b] \le T_{f_n - f_0}[a, b] + T_{g_n - g_0}[a, b] \rightarrow 0$$

and

$$T_{f_n g_n - f_0 g_0}[a, b] \le M \cdot T_{f_n - f_0}[a, b] + N \cdot T_{g_n - g_0}[a, b] \rightarrow 0,$$

where M is a uniform upper bound for $|g_n|$ and N an upper bound for $|f_0|$. The conclusions then follow from the previous theorem. $\square$

Further results on convergence in variation and convergence in length may be found in [M3].

§9.3. Norms on *AC*

We have previously observed that the set of absolutely continuous functions is a linear space. The purpose of this section is to show that, with a suitably defined norm, this space is a Banach space.

Definition 9.3.1 [G2]. Consider a function f which is summable on the interval $[0, 1]$. For each $h \in (0, 1)$, let

$$f_h(x) = \frac{1}{h} \int_0^h f(x + t) \, dt.$$

Clearly, f_h is defined on $[0, 1 - h]$. Extend f_h to $[0, 1]$ by defining it to be constant on $[1 - h, 1]$.

Lemma 9.3.2 [G2]. *For every function f which is summable on $[0, 1]$, the functions f_h are absolutely continuous.*

Proof. Let

$$F(x) = \int_0^x f(t) \, dt.$$

For pairwise disjoint intervals $[a_j, b_j], j = 1, \ldots, n$ in $[0, 1 - h]$, we have

$$\sum_{j=1}^n |f_h(b_j) - f_h(a_j)| \le \frac{1}{h} \sum_{j=1}^n [|F(b_j + h) - F(a_j + h)| + |F(b_j) - F(a_j)|].$$

It follows that f_h is absolutely continuous. $\square$

Lemma 9.3.3 [G2]. *Suppose $\{f_n\}$ is a sequence of absolutely continuous functions and $f_n \to f$ in the L^1 norm. If the sequence $\{f_n'\}$ of derivatives is a Cauchy sequence in L^1, then f is equivalent to an absolutely continuous function and $f_n' \to f'$ in the L^1 norm.*

Proof. Since $\{f_n'\}$ is Cauchy in L^1 (a complete space), there exists a function $g \in L^1$ such that $f_n' \to g$ in the L^1 norm. Since each f_n is absolutely continuous, we have

$$f_n(x) - f_n(0) = \int_0^x f_n'(t)\,dt.$$

Let

$$F(x) = \int_0^x g(t)\,dt.$$

Because $f_n' \to g$ in L_1, we have

$$F(x) = \lim_{n \to \infty} \{f_n(x) - f_n(0)\}$$

for every $x \in [0,1]$. Also, $f_n \to f$ in L^1; thus, there exists a subsequence of $\{f_n\}$ which converges a.e. to f. Hence, $f = F + c$ a.e. for some constant c. Finally, $F' = g$ a.e. Therefore, F' is the limit of the sequence $\{f_n'\}$ in the L^1 norm. □

Definition 9.3.4 [G2]. For any function f which is summable on $[0,1]$, let $V(f)$ denote the infimum of the variations on $[0,1]$ of functions which are equivalent to f. We observe that if f is absolutely continuous, then

$$V(f) = \int_0^1 |f'(t)|\,dt.$$

Lemma 9.3.5 [G2]. *For any summable f on $[0,1]$ and any $h \in (0,1)$, we have $V(f_h) \le V(f)$.*

Proof. By the fundamental theorem of calculus,

$$f_h'(x) = \frac{f(x+h) - f(x)}{h}$$

a.e. on $[0, 1-h]$. Then

$$V(f_h) = \int_0^{1-h} |f_h'(t)|\,dt = \frac{1}{h} \int_0^{1-h} |f(t+h) - f(t)|\,dt$$

$$= \frac{1}{h} \int_0^h \sum_{i=1}^n |f(t+ih) - f(t+(i-1)h)|\,dt,$$

where $(n-1)h \le 1 - h$ and $nh > 1 - h$. For every t, the integrand $\le V(f)$; thus, $V(f_h) \le V(f)$. □

Lemma 9.3.6 [G2]. *If f is absolutely continuous, then f_h' converges to f' in the L^1 norm as $h \to 0$.*

Proof. Since f is absolutely continuous,

$$V(f) = \int_0^1 |f'(x)| \, dx$$

and

$$V(f_h) = \frac{1}{h} \int_0^{1-h} |f(x + h) - f(x)| \, dx$$

for every $h \in (0, 1)$. Let $\varepsilon > 0$. There exists $\delta > 0$ such that $mE < \delta$ implies

$$\int_E |f'(x)| \, dx < \varepsilon.$$

Also,

$$f_h'(x) = \frac{f(x + h) - f(x)}{h} \to f'(x)$$

at every point of differentiability of f, i.e., a.e. Applying Egoroff's theorem, for every sequence of values of h converging to zero, there exists a set $T \subset [0, 1]$ such that $mT > 1 - \delta$ and $f_h' \to f'$ uniformly on T. Further, there exists $h_0 > 0$ such that if $0 < h < h_0$, then

$$\int_T |f_h'(x) - f'(x)| \, dx < \varepsilon,$$

as long as the values of h are in the sequence.

Now

$$\int_{[0,1]-T} |f_h'(x)| \, dx + \int_T |f_h'(x)| \, dx \le \int_{[0,1]-T} |f'(x)| \, dx + \int_T |f'(x)| \, dx,$$

so that

$$\int_{[0,1]-T} |f_h'(x)| \, dx \le \varepsilon + \int_T |f'(x)| \, dx - \int_T |f_h'(x)| \, dx$$

$$\le \varepsilon + \int_T |f'(x) - f_h'(x)| \, dx < 2\varepsilon.$$

Hence,

$$\int_0^1 |f'(x) - f_h'(x)| \, dx$$

$$\le \int_T |f'(x) - f_h'(x)| \, dx + \int_{[0,1]-T} |f'(x)| \, dx + \int_{[0,1]-T} |f_h'(x)| \, dx \le 4\varepsilon.$$

Since the above holds for every sequence of values of h which converge to zero, this implies that $f'_h \to f'$ in the L^1 norm. $\qquad\square$

Lemma 9.3.7 [G2]. *If f is summable on $[0, 1]$, then $V(f_h)$ converges to a finite number or to $+\infty$ as $h \to 0$. Further, $V(f_h)$ converges to a finite number if and only if f is equivalent to a function of bounded variation.*

Proof. Suppose f is of bounded variation. For each $h \in (0, 1)$, by Lemma 9.3.5,

$$V(f_h) = \frac{1}{h} \int_0^h \sum_{i=1}^n |f(t + ih) - f(t + (i - 1)h)| \, dt.$$

Let $\varepsilon > 0$. There is an $h_0 > 0$ such that for all $h \in (0, h_0)$, the integrand $> V(f) - \varepsilon$. Thus, $V(f_h) > V(f) - \varepsilon$. The result follows by Lemma 9.3.5.

Conversely, suppose f is not equivalent to any function of bounded variation. Then for each M, there exists an $h_0 > 0$ such that for every $h \in (0, h_0)$, we have $V(f_n) > M$; the result follows. $\qquad\square$

Theorem 9.3.8 [G2]. *Let $AC[0, 1]$ be the space of absolutely continuous function with the norm*

$$\|f\| = \|f\|_1 + \|f'\|_1.$$

The space $AC[0, 1]$ is a Banach space.

Proof. That $AC[0, 1]$ is linear was observed in Chapter 7. That it is complete was proved by Lemma 9.3.3. $\qquad\square$

Theorem 9.3.9 [G2]. *If $f \in L^1$, then $f_h \in AC$ for every $h \in (0, 1)$ and $\lim_{h \to 0} \|f_h\|$ exists (may be $+\infty$). Also, $\lim_{h \to 0} \|f_h\| < \infty$ if and only if f is equivalent to a function of bounded variation. f_h converges in AC, as $h \to 0$, if and only if f is equivalent to an absolutely continuous function and then f_h converges to f.*

Proof. Use Lemmas 9.3.3, 9.3.5, and 9.3.7 and the fact that for $f \in L^1$, $f_h \to f$ in L^1 as $h \to 0$. $\qquad\square$

Remark [G2]. The previous theorem has a geometric interpretation:

$\{f_h\}$ converges to the boundary of a ball, with center at the origin, in AC if and only if f is equivalent to a function of bounded variation;

$\{f_h\}$ converges to a point in AC if and only if f is equivalent to an absolutely continuous function.

Definition 9.3.10 [G4]. Let $I = [0, 1]$ and let $f : [0, 1] \to \mathbf{R}^1$ be of bounded variation. We define a measure α_f on the Borel sets in I. For each $x \in I$, let J_x be the interval whose endpoints are the right and left limits of $f(\xi)$ as $\xi \to x$.

Because f is continuous except, at most, on a countable set, J_x is a single point except for a countable set of values of x. For each Borel set E, $\alpha_f(E)$ is the one-dimensional Hausdorff measure of the planar set

$$\bigcup_{x \in E} J_x.$$

For the case where f is absolutely continuous, we have

$$\alpha_f(E) = \int_E \{1 + [f'(x)]^2\}^{1/2} \, dx.$$

We may also obtain the measure α_f as follows. Since there is a right continuous function which is equivalent to f, we will assume that f is right continuous. Let μ_f be the measure which is the derivative of f in the distribution sense. For each Borel set $E \subset I$, let

$$\alpha_f(E) = \sup \sum_{i=1}^{n} \{[\mu_f(E_i)]^2 + [mE_i]^2\}^{1/2},$$

where the supremum is taken over all partitions of E into finitely many pairwise disjoint Borel sets.

Lemma 9.3.11 [G4]. *If $f: I \to \mathbf{R}^1$ is differentiable a.e., then for every $\varepsilon > 0$, there exists a function $g: I \to \mathbf{R}^1$ which is of class C^1, such that $f(x) = g(x)$ except on a set of measure less than ε.*

Lemma 9.3.12 [G4]. *Given two points (a, c) and (b, d), with $a < b$, let L be the distance between them; let α and β be real numbers. For every $\varepsilon > 0$, there is a function $f: [a, b] \to \mathbf{R}^1$ which is of class C^1, with the left derivative at a equal to α and the right derivative equal to β such that $\alpha_f([a, b]) < L + \varepsilon$.*

Theorem 9.3.13 [G4]. *A function $f: I \to \mathbf{R}^1$ which is of bounded variation is equivalent to an absolutely continuous function if and only if, for every $\varepsilon > 0$, there is a function $g \in C^1$ such that if $G = \{x: f(x) \neq g(x)\}$, then $\alpha_f(G) < \varepsilon$ and $\alpha_g(G) < \varepsilon$.*

Proof. Suppose f satisfies the stated condition. Since $f \in BV$,

$$\int_I \sqrt{1 + [f'(x)]^2} \, dx < \infty.$$

Let $\varepsilon > 0$ and let $g \in C^1$ be such that if $G = \{x: f(x) \neq g(x)\}$, then $\alpha_f(G) < \varepsilon$ and $\alpha_g(G) < \varepsilon$. We have

$$\alpha_f(I) = \alpha_f(G) + \alpha_f(I - G) = \alpha_f(G) + \alpha_g(I - G) \leq \int_{I-G} \sqrt{1 + [f'(x)]^2} \, dx + \varepsilon$$

Since $f' = g'$ at almost every point of the set where $f = g$. Thus,

$$\alpha_f(I) \le \int_I \{1 + [f'(x)]^2\}^{1/2}\, dx$$

so that f is absolutely continuous.

Conversely, suppose f is absolutely continuous. Let $\varepsilon > 0$; there is a $\delta > 0$ such that for every Borel set E with $mE < \delta$, we have

$$\alpha_f(E) = \int_E \{1 + [f'(x)]^2\}^{1/2}\, dx < \frac{\varepsilon}{3}.$$

By Lemma 9.3.11, there is a function $h \in C^1$ such that $f(x) = h(x)$ except on a set E with $mE < \delta$. By adding a countable set to E, we have $A = I - E$ is perfect. Then E is the union of pairwise disjoint nonabutting open intervals $I_1, I_2, \dots$. Since $h \in C^1$, the series

$$\sum_{n=1}^{\infty} \alpha_h(I_n)$$

converges. Choose m so that

$$\sum_{n=m+1}^{\infty} \alpha_h(I_n) < \frac{\varepsilon}{3}.$$

Now $\alpha_f(E) < \varepsilon/3$ but $\alpha_h(E)$ may be too large, so we modify h on the finite set of intervals $I_1, \dots, I_m$ to obtain the desired g. By Lemma 9.3.12, we can define g on these intervals so that

$$\sum_{n=1}^{m} \alpha_g(I_n) < \sum_{n=1}^{m} \alpha_f(I_n) + \frac{\varepsilon}{3} \le \alpha_f(E) + \frac{\varepsilon}{3}.$$

Then

$$\alpha_g(E) = \sum_{n=m+1}^{\infty} \alpha_h(I_n) + \sum_{n=1}^{m} \alpha_g(I_n) < \frac{\varepsilon}{3} + \alpha_f(E) + \frac{\varepsilon}{3} < \varepsilon. \qquad \square$$

Theorem 9.3.14 [G4]. *If $f: I \to \mathbf{R}^1$ is of bounded variation, then for every $\varepsilon > 0$ there is a function $g: I \to \mathbf{R}^1$ which is of class C^1 such that $f(x) = g(x)$ except of a set of measure less than ε and $|\alpha_f(I) - \alpha_g(I)| < \varepsilon$.*

Proof. Suppose $f: I \to \mathbf{R}^1$ is of bounded variation and right continuous. By Lemma 9.3.11, there is a function $u \in C^1$ such that $f(x) = u(x)$ except on a set E of Lebesgue measure less than ε. We choose E so that $A = I - E$ is perfect and f is continuous at each point of A. The set E is the union of pairwise disjoint nonabutting open intervals of $I_1, I_2, \dots$. As in the proof of the previous theorem, there exists m such that

$$\sum_{n=m+1}^{\infty} \alpha_u(I_n) < \frac{\varepsilon}{2}.$$

By Lemma 9.3.12, we may modify u on the intervals $I_1, \ldots, I_m$ to obtain $v \in C^1$ such that

$$\sum_{n=1}^{m} \alpha_v(I_n) < \sum_{n=1}^{m} \alpha_f(I_n) + \frac{\varepsilon}{2}.$$

Then $f(x) = v(x)$ except on a set E of measure less than ε and $\alpha_v(I) < \alpha_f(I) + \varepsilon$. Finally, we modify v on I_1 to obtain a longer $g \in C^1$, but only long enough to have

$$\alpha_f(I) - \varepsilon < \alpha_g(I) < \alpha_f(I) + \varepsilon. \qquad \square$$

Definition 9.3.15 [GP, p. 156]. A function $\sigma: (0, \infty) \to \mathbf{R}^1$ is called a *modulus of continuity* if

$$\lim_{x \to 0} \sigma(x) = 0.$$

A function $f \in C[0, 1]$ is said to have σ as its modulus of continuity ($f \in C_\sigma$) if

$$|f(x) - f(y)| \leq \sigma(|x - y|)$$

for all $x, y \in [0, 1]$.

Lemma 9.3.16 [GP, p. 156]. *Let σ be a modulus of continuity. Let $0 < \varepsilon < 1$ and n be a positive integer. Then there exist δ and η, with $0 < \delta < 1/n$, $\eta > 0$, and a continuous function f, with $\|f\| \leq \varepsilon$ (see §6.3 for the definition of the norm on $C[0, 1]$), such that for every $x \in [0, 1 - 1/n]$ and every g, with $\|f - g\| \leq \eta$, we have either*

$$|g(x) - g(y)| > \sigma(|x - y|) \quad \text{for every } y \in \left[x - \frac{\delta}{2}, x - \frac{\delta}{4}\right]$$

or

$$|g(x) - g(y)| > \sigma(|x - y|) \quad \text{for every } y \in \left[x + \frac{\delta}{4}, x + \frac{\delta}{2}\right].$$

Proof. Let $\delta < 1/n$ so that $\sigma(\delta) < \varepsilon/8$. Let f be a continuous function whose graph is composed of a finite set of straight line segments such that $f(0) = 0$, $0 \leq f(x) \leq c$ for all $x \in [0, 1]$, and such that the line segments which form the graph of f have slopes $\pm \varepsilon/\delta$, and all but the last has length equal to $(\varepsilon^2 + \delta^2)^{1/2}$. There is a positive integer k such that $k\delta < 1 \leq (k + 1)\delta$. Then $[0, 1 - 1/n] \subset [0, k\delta]$.

For every $x \in [0, k\delta]$, $(y, f(y))$ is on the same line segment of the graph of f as is $(x, f(x))$ either for all $y \in [x - \delta/2, x - \delta/4]$ or for all $y \in [x + \delta/4, x + \delta/2]$. For all such y,

$$|f(x) - f(y)| = \frac{\varepsilon}{\delta}|x - y| > 8\sigma(|x - y|)\frac{|x - y|}{\delta} \geq 2\sigma(|x - y|),$$

since $\sigma(|x - y|) < \sigma(\delta) < \varepsilon/8$ and $|x - y| \geq \delta/4$. Let

$$\eta = \frac{1}{2}\sigma\left(\frac{\delta}{4}\right).$$

Suppose $\|f - g\| < \eta$. Then for x and y as above, we have

$$|g(x) - g(y)| > |f(x) - f(y)| - 2\eta$$

$$> 2\sigma(|x - y|) - \sigma\left(\frac{\delta}{4}\right)$$

$$\geq 2\sigma(|x - y|) - \sigma(|x - y|) = \sigma(|x - y|),$$

since $\sigma(|x - y|) \geq \sigma(\delta/4)$. $\qquad\square$

Lemma 9.3.17 [GP, p. 157]. *Let σ be a modulus of continuity. For every positive integer n, the set $E_n \subset C[0,1]$ of functions f for which there is a δ, with $0 < \delta < 1/n$, depending on f, such that for every $x \in [0, 1 - 1/n]$,*

$$|f(x) - f(y)| > \sigma(|x - y|)$$

either for every $y \in [x - \delta/2, x - \delta/4]$ or for every $y \in [x + \delta/4, x + \delta/2]$ contains a dense open subset in $C[0,1]$.

Proof. Let $g \in C[0,1]$ and $\zeta \in (0,1)$. There is a polynomial p such that $\|p - g\| < \zeta/2$. Let $M = \max\{|p'(x)|: x \in [0,1]\}$. For ε and σ of the previous lemma, take $\zeta/2$ and the function τ defined by $\tau(x) = \sigma(x) + Mx$. Then there exist δ and η, with $0 < \delta < 1/n$ and $\eta > 0$, and a continuous function h, with $\|h\| \leq \zeta/2$, such that $\|k - h\| < \eta$ implies that, for every $x \in [0, 1 - 1/n]$,

$$|k(x) - k(y)| > \tau(|x - y|)$$

either for every $y \in [x - \delta/2, x - \delta/4]$ or for every $y \in [x + \delta/4, x + \delta/2]$. Then, for all such x and y, we have

$$|(p + k)(x) - (p + k)(y)| \geq |k(x) - k(y)| - |p(x) - p(y)|$$

$$> \tau(|x - y|) - M|x - y|$$

$$= \sigma(|x - y|).$$

Since

$$\|g - (p + h)\| \leq \|g - p\| + \|h\| < \zeta,$$

the sphere of center g and radius ζ contains an open subset of E_n. Since $g \in C[0,1]$ and ζ are arbitrary, E_n contains a dense open subset of $C[0,1]$. $\qquad\square$

Theorem 9.3.18 [GP, p. 158]. *For every modulus of continuity σ, there is an $f \in C[0,1]$ such that, for every $g \in C_\sigma$, $f(x) \neq g(x)$ a.e. Moreover, the set of functions which do not have this property is of the first category.*

Proof. Let E_n, $n = 1, 2, \ldots$, be as defined in the previous lemma and let

$$E = \bigcap_{n=1}^{\infty} E_n.$$

Then the set E is residual in $C[0,1]$; i.e., its complement is of the first category in $C[0,1]$.

Let $f \in E$. Then for every n, there is a $\delta_n \in (0, 1/n)$, such that for every $x \in [0, 1 - 1/n]$, we have

$$|f(x) - f(y)| > \sigma(|x - y|)$$

either for every $y \in [x - \delta_n/2, x - \delta_n/4]$ or for every $y \in [x + \delta_n/4, x + \delta_n/2]$. Now suppose $g \in C_\sigma$ and suppose the set

$$A = \{x : f(x) = g(x)\}$$

has positive measure. Let $x \in (0, 1)$ be a point in A at which the density of A is 1. Then the set

$$B = \{y : |f(x) - f(y)| > \sigma(|x - y|\}$$

has density 0 at x. Choose n_0 so that $x < 1 - 1/n_0$. Then for every $n \geq n_0$, since $f \in E_n$,

$$|f(x) - f(y)| > \sigma(|x - y|)$$

on a subset of $[x - \delta_n/2, x + \delta_n/2]$ of relative measure $\frac{1}{4}$. Since

$$\lim_{n \to \infty} \delta_n = 0,$$

this contradicts the assertion that B has density 0 at x.   □

Theorem 9.3.19 [G3]. *For every modulus of continuity σ, there is an $f \in C^1[0,1]$ such that, for every $g \in C^1[0,1]$ with $g' \in C_\sigma$, $f(x) \neq g(x)$ a.e.*

Proof. Let $h \in C[0,1]$ be such that, for every $k \in C_\sigma$, $h(x) \neq g(x)$ a.e. Let

$$f(x) = \int_0^x h(t)\, dt, \quad 0 \leq x \leq 1.$$

Suppose there is a differentiable function g such that $g' \in C_\sigma$ and $f(x) = g(x)$ on a set E for which $mE > 0$. Almost every point of E is a point of density 1 of E. Clearly, for every such x, $f'(x) = g'(x)$. Hence, the set of points for which $f'(x) = g'(x)$ has positive measure. But $f'(x) = h(x)$ for every x. Since $g' \in C_\sigma$, this contradicts the assumption about h.   □

Corollary 9.3.20 [G3]. *For every modulus of continuity σ, there is an absolutely continuous function f such that, for every $g' \in C^1[0,1]$ with $g' \in C_\sigma$, $f(x) \neq g(x)$ a.e.*

Remark [G5]. These results suggest the following metric for the space of *AC* functions. For $f, g \in AC$, let

$$d(f,g) = mE + \int_E |f'(t)|\, dt + \int_E |g'(t)|\, dt,$$

where $E = \{x: f(x) \neq g(x)\}$. We leave it to the reader to show that *AC* is a Banach space with this metric.

§9.4. Norms on *BV*

The results of this section are found in [A1].

We recall the norm defined on $BV[a, b]$ in §6.3: If $f \in BV[a, b]$,

$$\|f\| = |f(a)| + T_f[a, b].$$

In that section, we showed that with such a norm $BV[a, b]$ is a Banach space; so also are the subspaces $CBV[a, b]$ (continuous functions of bounded variation) and *AC*. With this metric *AC* is a separable space, although neither *BV* nor *CBV* is. Indeed, every uncountable set in *BV* is nondense as illustrated by the following example.

EXAMPLE 9.4.1. Let x_h $(0 < h < 1)$ be the characteristic function of the point h. Then for $h_1 \neq h_2$, we have $\|x_{h_1} - x_{h_2}\| = T_{x_{h_1} - x_{h_2}}[0, 1] = 4$; and there exists a continuum of disjoint spheres $K(x_h, 1) \subset K(0, 3)$ (where 0 denotes the zero element of the space *BV*). Hence, an arbitrary sphere $K(x, r) \subset BV$ contains a continuum of disjoint spheres $K(rx_h/3 + x, r/3)$.

We turn our attention now to some other metrics which have been defined on *BV* in such a manner that convergence in the metric will be closely related to convergence in variation or convergence in length.

Definition 9.4.2. Let $x, y \in BV[0, 1]$. We define the following four metrics:

$$\|x - y\|_1 = \int_0^1 |x(t) - y(t)|\, dt + |T_x[0, 1] - T_y[0, 1]|,$$

$$\|x - y\|_2 = \int_0^1 |x(t) - y(t)|\, dt + |L_x[0, 1] - L_y[0, 1]|,$$

$$\|x - y\|_3 = \sup_{0 \leq t \leq 1} |x(t) - y(t)| + |T_x[0, 1] - T_y[0, 1]|,$$

$$\|x - y\|_4 = \sup_{0 \leq t \leq 1} |x(t) - y(t)| + |L_x[0, 1] - L_y[0, 1]|.$$

Clearly, $\|x - y\|_1 \leq \|x - y\|_3$ and $\|x - y\|_2 \leq \|x - y\|_4$ for every $x, y \in BV[0, 1]$. With the metric $\|\cdot\|_3$ (or $\|\cdot\|_4$) each point of the set of discontinuous functions of bounded variation (*DBV*) is the center of a sphere containing no points of *DBV*. This is contrast to the fact that a sequence of continuous functions may converge in variation (or length) to a discontinuous function; see Corollary 9.1.7 and Corollary 9.2.13.

Theorem 9.4.3.

(a) $x_n - v \to x \Rightarrow \|x_n - x\|_1 \to 0$.
(b) *If* x *is continuous*, $\|x_n - x\|_1 \to 0 \Rightarrow x_n - v \to x$.
(c) $x_n - L \to x \Rightarrow \|x_n - x\|_2 \to 0$.
(d) *If* x *is continuous*, $\|x_n - x\|_2 \to 0 \Rightarrow x_n - L \to x$.
(e) *If* x *is continuous*,

$$\|x_n - x\|_2 \to 0 \Rightarrow x_n - L \to x$$

$$\Rightarrow x_n - v \to x$$

$$\Rightarrow \|x_n - x\|_1 \to 0.$$

Proof. The proofs of these statements are left to the reader; they follow readily from the results of §§9.1 and 9.2. □

Remark. Part (b) may fail if we remove the condition of continuity of the limit function x. Adams [A1] remarks that if x is discontinuous at even a single point, then x_n may fail to converge to x at every point of $[0, 1]$. As an example, he suggests modifying a sequence which converges in the mean without converging anywhere. If x has even a single discontinuity, then $\|x_n - x\|_2 \to 0$ may fail to imply that $\|x_n - x\|_1 \to 0$. Consider Example 9.1.13; modify the limit function f_0 so that it has a suitable jump at $t = 1$. From Theorem 9.2.10, we see that the converse of (e) does not hold, i.e., $\|x_n - x\|_1 \to 0$ does not imply $\|x_n - x\|_2 \to 0$ even if all functions are *AC*.

With either of the metrics $\|\cdot\|_3$ or $\|\cdot\|_4$, the spaces *BV*, *CBV*, and *AC* are not complete or even locally compact. Consider a sequence of polygonal functions all having the same total variation and the same length on $[0, 1]$ and converging uniformly to the zero function to verify this.

Theorem 9.4.4. *With either of the metrics* $\|\cdot\|_1$ *or* $\|\cdot\|_2$, *every bounded portion of BV is compact.*

Proof. Let $\{x\}$ be a set of functions satisfying the condition

$$\|x - 0\|_1 = \int_0^1 |x(t)|\, dt + T_x[0, 1] \le M.$$

Such a set is obviously of uniformly bounded variation and hence uniformly bounded. By Helly's First Theorem (6.1.18), this set contains a sequence x_n converging everywhere to a function $x \in BV$. By Lebesgue's convergence theorem, x_n converges in the mean to x. Since the sequence $T_{x_n}[0, 1]$ is bounded, it contains a subsequence $T_{x_{n_i}}[0,1]$ which converges to a limit τ. By Theorem 9.1.3, we have

$$T_x[0,1] \leq \varliminf T_{x_{n_i}}[0,1].$$

Thus $T_x[0,1] \leq \tau$. If necessary, we may redefine $x(0)$ so that $T_x[0,1] = \tau$. Then $\|x_{n_i} - x\|_1 \to 0$, and the compactness is proved. Hence, the space *BV* with the metric $\|\cdot\|_1$ is complete. Using the same argument as above with length instead of total variation, the same result can be proved for the $\|\cdot\|_2$ metric. $\square$

Theorem 9.4.5. *With either metric* $\|\cdot\|_1$ *or* $\|\cdot\|_2$, *the space BV is separable.*

Proof. We will show that there exists a denumerable subset of *AC* which is dense in *BV*. By Theorem 9.2.9, any function $x \in BV$ can be approximated in either metric by an inscribed polygonal function y. Any such function can be approximated in the metric by a continuous polygonal function z whose vertices have rational coordinates. Such functions z are a denumerable subset of *AC*. $\square$

Theorem 9.4.6. *For every real number a and every x, y ∈ BV,* $\|ax - ay\|_1 = |a| \cdot \|x - y\|_1$.

Proof. The proof is left to the reader. Note that this theorem says that the metric $\|\cdot\|_1$, considered as a functional defined on the space of *BV* functions, is *homogeneous*. $\square$

The last three results show that $\|\cdot\|_1$ is the most desirable of the four metrics we have considered. Thus, for the remainder of this section, this metric will be employed for *BV*.

Definition 9.4.7. Suppose $U: E_1 \to E_2$, E_1 and E_2 are metric spaces. If U satisfies a Lipschitz condition, i.e., there exists a number M such that

$$\|U(x) - U(y)\|_{E_2} \leq M \cdot \|x - y\|_{E_1},$$

for every $x, y \in E_1$, then the smallest number M for which the above holds is called the *modulus of* $U(x)$ *on* E_1 and designated by $\mathrm{mod}_{E_1} U$.

Note that if U is homogeneous, then for every real number a and every $x, y \in E_1$, we have

$$\frac{\|U(ax) - U(ay)\|_{E_2}}{\|ax - ay\|_{E_1}} = \frac{\|aU(x) - aU(y)\|_{E_2}}{|a| \cdot \|x - y\|_{E_1}} = \frac{\|U(x) - U(y)\|_{E_2}}{\|x - y\|_{E_1}},$$

from which it follows that

(9.4.8) $$\mathrm{mod}_{E_1} U = \sup_{x,y \in E_1, \|x-y\|_{E_1}=1} \|U(x) - U(y)\|_{E_2}.$$

Theorem 9.4.9. *If α is any summable function, the integral*

(9.4.10)
$$\int_0^1 x(t)\alpha(t)\,dt$$

defines on the space BV a continuous linear functional f. The functionals generated by α_1 and α_2 are identical if and only if $\alpha_1(t) = \alpha_2(t)$ a.e. on $[0, 1]$. The functional f defined by (9.4.10) is uniformly continuous on BV if and only if α is essentially bounded on $[0, 1]$; in that case we have

$$\operatorname{mod}_{BV} f = \operatorname*{ess\,sup}_{0 \le t \le 1} |a(t)|.$$

Proof. The existence and linearity of f are clear. To prove the continuity, let $x \in BV$ and let $\{x_n\}$ be any sequence in BV such that $\|x_n - x\|_1 \to 0$. Then there exists a number K' such that for all n

$$\|x_n - x\|_1 = \int_0^1 |x_n - x|\,dt + |T_{x_n}[0,1] - T_x[0,1]| \le K'.$$

Hence, for each n there is at least one point $\bar{t}_n$ such that

$$|x_n(\bar{t}_n) - x(\bar{t}_n)| \le K'.$$

But for each n and all t, we have

$$|x_n(t) - x_n(\bar{t}_n)| \le T_{x_n}[0,1],$$
$$|x(t) - x(\bar{t}_n)| \le T_x[0,1],$$
$$T_{x_n}[0,1] \le T_x[0,1] + K'.$$

Thus, for all n and all t,

$$|x_n(t) - x(t)| \le 2T_x[0,1] + 2K' = K.$$

Moreover, given $\varepsilon > 0$, there exists $\delta > 0$ such that

$$\int_E K|\alpha|\,dt < \varepsilon$$

whenever $mE < \delta$. Choosing N so that $m\{t: |a(t)| > N\} < \delta$, we have

$$|f(x_n) - f(x)| = |f(x_n - x)| \le \left[\int_E + \int_{E^0}\right]|x_n - x|\cdot|\alpha|\,dt$$

$$\le \varepsilon + N\int_0^1 |x_n - x|\,dt < 2\varepsilon$$

if n is sufficiently large.

It is clear that if $\alpha_1 = \alpha_2$ a.e., they generate the same functional. Conversely, if α_1 and α_2 generate the same functional, we can choose

$$x(t) = \begin{cases} 1, & 0 \le t \le y \\ 0, & y < t \le 1, \end{cases}$$

for $y \in [0, 1]$. Then

$$\int_0^y \alpha_1(t) \, dt = \int_0^y \alpha_2(t) \, dt$$

for all y. Thus, $\alpha_1 = \alpha_2$ a.e. This fact, along with the proof of the next theorem, shows that a uniformly continuous functional f cannot be generated by an α which is not essentially bounded.

That f is uniformly continuous when α is essentially bounded follows from (9.4.8); we have

$$\text{mod}_{BV} f = \sup_{x, y \in BV, \|x - y\|_1 = 1} |f(x) - f(y)| \leq \underset{0 \leq t \leq 1}{\text{ess sup}} |\alpha(t)| = B.$$

We will show that in this case $\text{mod}_{BV} f = B$. Let $\varepsilon > 0$ be arbitrarily small and let $mE_1 = k_1$ and $mE_2 = k_2$, where

$$E_1 = \{t : \alpha(t) > B - \varepsilon\},$$

$$E_2 = \{t : \alpha(t) < \varepsilon - B\}.$$

Then either $k_1 > 0$ or $k_2 > 0$. Since the cases are symmetrical, we will suppose that $k_1 > 0$. Let $\delta > 0$ be chosen such that $0 < \delta < k_1 \varepsilon$. We can find an open set $O_1 \supset E_1$ such that $k_1 \leq mO_1 \leq k_1 + \delta$. From O_1 select a finite number of open intervals whose union is O_2 such that $mO_2 \geq mO_1 - \delta \geq k_1 - \delta$. Define

$$\phi(t) = \begin{cases} \dfrac{1}{k_1}, & t \in O_2 \\ 0, & \text{elsewhere.} \end{cases}$$

Then $\phi \in BV$ and

$$\int_0^1 \phi(t)\alpha(t) \, dt \geq (k_1 - \delta)\frac{1}{k_1}(B - \varepsilon) - \delta \frac{1}{k_1} B \geq B - \varepsilon - 2B\varepsilon,$$

$$\int_0^1 \phi(t) \, dt \leq \frac{1}{k_1}(k_1 + \delta) \leq 1 + \varepsilon.$$

Let $x(t) = -y(t) = \phi(t)/2$ for all $t \in [0, 1]$. Then $T_x[0, 1] = T_y[0, 1]$ and

$$|f(x) - f(y)| = \left| \int_0^1 (x - y)\alpha \, dt \right| = \int_0^1 \phi(t)\alpha(t) \, dt \geq B - \varepsilon - 2B\varepsilon,$$

$$\|x - y\|_1 = \int_0^1 |x - y| \, dt = \int_0^1 \phi(t) \, dt \leq 1 + \varepsilon.$$

Thus,

$$\frac{|f(x) - f(y)|}{\|x - y\|_1} \geq \frac{B - \varepsilon - 2B\varepsilon}{1 + \varepsilon}.$$

Since the right side of the above inequality is arbitrarily close to B, we conclude that $\mathrm{mod}_{BV} f = B$.   □

Theorem 9.4.11. *Every additive and uniformly continuous functional defined on* *BV* *can be expressed in the form* (9.4.10), *with* α *summable, bounded, and independent of* x. *Moreover, we have*

$$\|f\|_{BV} \le \int_0^1 |\alpha(t)| \, dt \le \mathrm{mod}_{BV} f = B,$$

where $B = \mathrm{ess\ sup}_{0 \le t \le 1} |\alpha(t)|$.

Proof. Let f be any additive and uniformly continuous functional defined on *BV*. Let

$$\xi_t(u) = \begin{cases} 1, & 0 \le u \le t \\ 0, & t < u \le 1, \end{cases}$$

and $f(\xi_t) = g(t)$. For $0 \le t_1 < t_2 \le 1$, we have

(9.4.12)

$$|g(t_1) - g(t_2)| = |f(\xi_{t_1}) - f(\xi_{t_2})|$$

$$\le M \cdot \|\xi_{t_1} - \xi_{t_2}\|_1 = M \int_0^1 (\xi_{t_1} - \xi_{t_2}) \, dt + M \, |T_{\xi_{t_1}}[0,1] - T_{\xi_{t_1}}[0,1]|,$$

where $M = \mathrm{mod}_{BV} f$. For $t_2 < 1$, the second term on the right vanishes, and we have

$$|g(t_1) - g(t_2)| \le M(t_2 - t_1);$$

i.e., g satisfies a Lipschitz condition on $[0,1)$ and is absolutely continuous there. Now define

$$\bar{g}(t) = \begin{cases} g(t) & 0 \le t < 1 \\ g(1-), & t = 1. \end{cases}$$

The existence of the Stieltjes integral

$$\int_0^1 x(t) \, d\bar{g}(t)$$

is then assured for every $x \in BV$.

Let x be any function in *BV* vanishing at $t = 0$. By Theorem 9.2.9, there exists a sequence of polygonal functions p_n $(n = 1, 2, 3, \ldots)$ inscribed in x such that $p_n - v \to x$. For each n, the vertices of the functions p_n are determined by a set P_n of points

$$0 = t_{0,n} < t_{1,n} < t_{2,n} < \cdots < t_{r(n),n} = 1$$

with $P_n \subset P_{n+1}$ for every n and

$$P = \bigcup_{n=1}^{\infty} P_n$$

everywhere dense in $[0, 1]$. We now define a step function z_n associated with each p_n:

$$z_n(t) = \begin{cases} 0, & t = 0 \\ x(t_{i,n}), & t_{i-1,n} < t \le t_{i,n}, \end{cases} \quad \text{all } i \text{ and } n.$$

Then $T_{z_n}[0, 1] = T_{p_n}[0, 1]$; therefore, $T_{z_n}[0, 1] \to T_x[0, 1]$ as $n \to \infty$. Also, since x is Riemann integrable on $[0, 1]$, z_n converges in the mean to x. Hence, $\|z_n - x\|_1 \to 0$ which implies $f(z_n) \to f(x)$.

Note $z_n(u)$ may be expressed in the form

$$z_n(u) = \sum_i x(t_{i,n})[\xi_{t_{i,n}}(u) - \xi_{t_{i-1,n}}(u)];$$

therefore, since f is additive, we must have

$$f(z_n) = \sum_i x(t_{i,n})[g(t_{i,n}) - g(t_{i-1,n})]$$

$$= \sum_i x(t_{i,n})[\bar{g}(t_{i,n}) - \bar{g}(t_{i-1,n})] + x(1)[g(1) - \bar{g}(1)].$$

Hence,

$$(9.4.13) \qquad f(x) = \lim_{n \to \infty} f(z_n) = \int_0^1 x(t)\, d\bar{g} + x(1)[g(1) - \bar{g}(1)],$$

which is the form of the functional for every x vanishing at $t = 0$. Any function $x \in BV$ not vanishing at $t = 0$ can be expressed as the sum $x_1 + x_2$ where

$$x_1(t) = \begin{cases} x(t), & 0 < t \le 1 \\ 0, & t = 0, \end{cases} \qquad x_2(t) = \begin{cases} x(0), & t = 0 \\ 0, & 0 < t \le 1. \end{cases}$$

Since f is additive, we have $f(x) = f(x_1) + f(x_2)$. Since f is continuous, it must assume the same value at any two points x_2 and x_3 such that $x_2 = x_3$ a.e. and have the same total variations. Choosing x_3 as

$$x_3 = \begin{cases} \dfrac{x(0)}{2}, & t = \dfrac{1}{2} \\ 0, & \text{otherwise,} \end{cases}$$

then $x_2 = x_3$ a.e. and $T_{x_2}[0, 1] = T_{x_3}[0, 1]$; $f(x_3)$ is given by (9.4.13) and has the value zero. Hence, $f(x)$ is given by (9.4.13) even when $x(0) \neq 0$. Next consider the functions

$$x_4(t) = \begin{cases} 0, & 0 \le t < 1 \\ 1, & t = 1, \end{cases} \qquad x_5(t) = \begin{cases} 0, & 0 \le t < 1 \\ -1, & t = 1. \end{cases}$$

Then $x_4 = x_5$ a.e. and $T_{x_4}[0, 1] = T_{x_5}[0, 1]$; therefore, we must have $f(x_4) = f(x_5)$. By (9.4.13), we have $g(1) - \bar{g}(1) = -[g(1) - \bar{g}(1)]$ which implies $g(1) = \bar{g}(1)$ and (9.4.13) reduces to

$$(9.4.14) \qquad\qquad f(x) = \int_0^1 x(t)\, dg$$

for all $x \in BV$, where g is absolutely continuous on $[0, 1]$. The form (9.4.10) for f is an immediate consequence, with α bounded, since g satisfies a Lipschitz condition on $[0, 1]$. The evaluation of $\mathrm{mod}_{BV} f$ was made in the previous theorem. $\square$

Now we estimate $\|f\|_{BV}$. Recall the definition (6.3.2) of the norm of a functional. From (9.4.14), we have

$$|f(x)| \le \left[\sup_{0 \le t \le 1} |x(t)|\right] T_g[0, 1]$$

and

$$\frac{|f(x)|}{\|x\|_1} \le \frac{\sup|x(t)|}{\int_0^1 |x(t)|\, dt + T_x[0, 1]} T_g[0, 1] \le T_g[0, 1],$$

from which we get the bound asserted in the theorem. Second, for $0 \le t_1 < t_2 \le 1$, we have

$$|g(t_1) - g(t_2)| = |f(\xi_{t_1} - \xi_{t_2})| \le \|f\|_{BV} \cdot \|\xi_{t_1} - \xi_{t_2}\|_1$$

$$= \|f\|_{BV} \cdot \left[\int_0^1 (\xi_{t_2} - \xi_{t_1})\, dt + T_{(\xi_{t_2} - \xi_{t_1})}[0, 1]\right],$$

which implies

$$\tfrac{1}{3}\omega[g, t] \le \|f\|_{BV}.$$
$$_{0 \le t \le 1}$$

Third, we note that for $x(t) \equiv 1$ for $t \in [0, 1]$, we have

$$f(x) = f(\xi_1) = g(1) = \int_0^1 dg = g(1) - g(0),$$

which implies that $g(0) = 0$. Finally, we make use of the relation

$$\int_0^1 x(t)\, dt = x(1)g(1) - \int_0^1 g(t)\, dx.$$

From this we have

$$\frac{|f(x)|}{\|x\|_1} = \frac{|f(x)|}{\int_0^1 |x(t)|\, dt + T_x[0, 1]}$$

$$\le |g(1)| \frac{|x(1)|}{\int_0^1 |x(t)|\, dt + T_x[0, 1]} + \max_{0 \le t \le 1} |g(t)|,$$

from which it follows that

(9.4.15) $$\|f\|_{BV} \leq |g(1)| + \max_{0 \leq t \leq 1} |g(t)|.$$

From the relations

$$|g(1)| = |f(\xi_1)| \leq \|f\|_{BV} \cdot \|\xi_1\|_1 = \|f\|_{BV},$$

$$|g(t)| = |f(\xi_t)| \leq \|f\|_{BV} \cdot \|\xi_t\|_1 \leq 2\|f\|_{BV},$$

we obtain

$$\max\left[|g(1)|, \tfrac{1}{2} \max_{0 \leq t \leq 1} |g(t)|\right] \leq \|f\|_{BV},$$

which implies

$$\|f\|_{BV} \geq \frac{K}{3},$$

where K is the right-hand member of (9.4.15).

Theorem 9.4.16. *Every additive and uniformly continuous functional f defined on the space CBV (AC) can be expressed in the form (9.4.10) where α is summable, bounded, and independent of x and $\mathrm{mod}_{BV} f = B$ as before.*

Proof. We have already noted that *CBV* is dense in *BV*. Let $x \in BV - CBV$ and $\{x_n\}$ any sequence in *CBV* such that $\|x_n - x\|_1 \to 0$. If f is uniformly continuous on *CBV*, $\lim f(x_n)$ exists and we may define $f(x)$ as this limit. Moreover, this limit is independent of the sequence x_n employed and f thus extended over *BV* is uniformly continuous. We now show that the extended functional is additive over *BV*. Let $x, y \in BV$, at least one of which is not in *CBV*. A sequence of partitions P_n of $[0,1]$ can be determined so that the polygonal functions x_n, y_n, and z_n, respectively, inscribed in x, y and $x + y$, whose vertices are determined by P_n, converge in the metric $\|\cdot\|_1$ to x, y, and $x + y$. We then have

$$z_n(t) = x_n(t) + y_n(t) \quad \text{for all } n \text{ and } t,$$

$$f(z_n) = f(x_n) + f(y_n),$$

$$\lim f(x_n) = f(x),$$

$$\lim f(y_n) = f(y),$$

$$\lim f(z_n) = f(x + y),$$

from which we get $f(x + y) = f(x) + f(y)$. That $\mathrm{mod}_{BV} f = B$ can be shown in the same manner as in the proof of Theorem 9.4.9 by appropriately smoothing of the function ϕ to make it continuous. A similar proof will hold for the space *AC*. $\qquad\square$

CHAPTER 10

Metric Separability

In Chapters 1–9, we have generally considered functions defined on intervals (possible infinite) of the real line. In this brief chapter, we extend some of our previous results to functions defined on arbitrary sets. The results of this chapter are based on [J1] and make use of the concept of metric separability. We begin with some definitions.

Definition 10.1. Let $|A|$ denote the outer measure of a set. If A is an interval, then $|A|$ is its length. If A is a single point or empty, $|A| = 0$.

Definition 10.2. Let $\mathscr{F}$ be the class of sets, each of which can be written as $\bigcup I_i$ where each I_i is an open, half-open, or closed interval or a single point and $I_i \cap I_j = \varnothing$ when $i \neq j$. If $A \in \mathscr{F}$, $|A| = \sum |I_i|$.

Remark. Clearly, there may be more than one way of expressing A as the union of sets of mutually disjoint intervals and points. In Exercise 1, the reader is asked to show that $|A|$ is independent of the representation of A.

Definition 10.3. If A is any set, $|A| = \inf\{|\alpha|: \alpha \text{ is open}, A \subset \alpha\}$.

Definition 10.4. The sets A and B are *metrically separated* if, for every $\varepsilon > 0$, there exist opens sets α and β such that $A \subset \alpha$ and $B \subset \beta$ with $|\alpha \cap \beta| < \varepsilon$.

Recall that for any set A, any point x (not necessarily in A), and I an interval containing x, then

$$\limsup_{|I| \to 0} \frac{|A \cap I|}{|I|}$$

is the *upper density* of A at the point x. The *lower density* of A at x is

$$\liminf_{|I| \to 0} \frac{|A \cap I|}{|I|}.$$

If these limits are equal, their common value is the *density* of A at x.

Theorem 10.5. *If A and B are metrically separated, then at almost all points of one set the density of the other set is zero.*

Proof. Let $\{\lambda_n\}$ be a decreasing sequence of positive numbers with $\lambda_n \to 0$. Let A_n be the points of A at which the upper density of B is greater than λ_n. Let $\varepsilon > 0$ be arbitrary. Then there exist open sets α and β such that $A \subset \alpha$, $B \subset \beta$, and $|\alpha \cap \beta| < \varepsilon$. With each $x \in A_n$, there is associated a sequence of closed intervals $\{I_n\}$ such that $x \in I_i \subset \alpha$ and

$$\frac{|B \cap I_i|}{|I_i|} > \lambda_n.$$

The intervals $\{I_n\}$ form a Vitali covering for the set A_n. Hence, there is a finite number $I_1, \ldots, I_k$ of these intervals which are mutually disjoint and

$$|A_n| + \varepsilon > \sum_{i=1}^{k} |I_i| \geq \sum_{i=1}^{k} |A_n \cap I_i| > |A_n| - \varepsilon.$$

Since $I_i \subset \alpha$ and $B \subset \beta$, we have

$$\varepsilon > |\alpha \cap \beta| > \sum_{i=1}^{k} |B \cap I_i| > \lambda_n \sum_{i=1}^{k} |I_i| > \lambda_n[|A_n| - \varepsilon].$$

Since ε is arbitrary, we get a contradiction unless $|A_n| = 0$.

The part of A at which the upper density of B is zero is the set

$$A_1 \cup A_2 \cup \cdots,$$

which is of measure zero, since each A_n is of measure zero. Since the lower density is less than the upper density, we obtain that at almost all points of A, the density of B is zero. In a similar manner, it can be shown that, at almost all points of B, the density of A is zero. $\qquad\square$

Theorem 10.6. *If at almost all points of a set A, the density of a set B is zero, then A and B are metrically separated.*

Proof. The proof is similar to the previous one and is left as an exercise for the reader. $\qquad\square$

Theorem 10.7. *A necessary and sufficient condition that two sets be metrically separated is that at almost all points of one set the density of the other set is zero.*

Proof. This follows from Theorems 10.5 and 10.6.

Theorem 10.8. *If the sets A and B are metrically separated, there exist measurable sets E and F with $A \subset E$ and $B \subset F$ and $|E \cap F| = 0$.*

Proof. Let $\{\varepsilon_n\}$ be a strictly decreasing sequence of positive numbers which converge to zero. Let α_n and β_n be open sets such that $A \subset \alpha_n$, $B \subset \beta_n$, and $|\alpha_n \cap \beta_n| < \varepsilon_n$. Let

$$E = \bigcap_{n=1}^{\infty} \alpha_n, \qquad F = \bigcap_{n=1}^{\infty} \beta_n.$$

Then E and F are measurable and, for every n, $E \cap F \subset \alpha_n \cap \beta_n$. Thus,

$$|E \cap F| < \varepsilon_n$$

for every n. Since $\varepsilon_n \to 0$, the result follows. $\square$

Theorem 10.9. *Let A be a measurable set and let $A = A_1 \cup A_2$, where A_1 and A_2 are metrically separated. Then A_1 and A_2 are measurable.*

Proof. By the previous theorem, there exist measurable sets E and F such that $A_1 \subset E$, $A_2 \subset F$, and $|E \cap F| = 0$. The set $A \cap E$ is measurable. Let $G = A \cap E - A_1$; thus, $A_1 \subset A \cap E = A_1 \cup G$. Then $G \subset A_2$, from which it follows that $G \subset E \cap F$. This gives

$$0 \leq |G| \leq |E \cap F| = 0.$$

Thus, G is measurable, which implies that $A_1 = A \cap E - G$ is measurable. Therefore, $A_2 = A - A_1$ is also measurable. $\square$

Definition 10.10. If f is defined on a set A, the ratio

$$\frac{f(x + h) - f(x)}{h}, \quad x \in A, x + h \in A,$$

may not tend to a limit as $h \to 0$. But if x is a limit point of A on the right and $x, x + h \in A$, then the ratios

$$\limsup_{h \to 0+} \frac{f(x + h) - f(x)}{h},$$

$$\liminf_{h \to 0+} \frac{f(x + h) - f(x)}{h}$$

both exist and are denoted by $_A D^+ f(x)$ and $_A D_+ f(x)$, respectively, and are called the *upper right and lower right derivates of f over A*. The *upper and lower left derivates*, $_A D^- f(x)$ and $_A D_- f(x)$, are similarly defined.

Theorem 10.11. *Let f be defined and measurable on the measurable set A. Then at all points of A, with the possible exception of a denumerable set, the four derivates of f over A exist and are measurable on A.*

Proof. If f is defined on a set A and $x \in A$ is a limit point of A on both sides, then all four derivates exist at x. A point $x \in A$ which is not a limit point of A on one side at least is an end point of an open interval of the set complementary to $\bar{A}$. Since this set of intervals is at most denumerable, it follows that the set of points of A which are not limit points of A on both sides is at most denumerable.

We will prove the measurability of $_A D^+ f(x)$, the proofs for the other derivates being similar. For any real number a, let

$$E^a = \{x \in A: {}_A D^+ f(x) < a\},$$

$$E_a = \{x \in A: {}_A D^+ f(x) \geq a\}.$$

If $_A D^+ f(x)$ is not measurable, then for some value of a the set E^a is not measurable. Since $A = E^a \cup E_a$ and A is measurable, then (by Theorem 10.9) E^a and E_a are not metrically separated. Let $\{c_n\}$ be a strictly increasing sequence of numbers with $c_n \to a$. Let E_{jk} be the points of E^a for which

$$(10.12) \qquad \frac{f(\xi + h) - f(\xi)}{h} < c_j, \quad \xi \in E^a,$$

when $0 < h < 1/k$, ξ, $\xi + h \in A$. If $j_2 \geq j_1$, $k_2 \geq k_1$ then $E_{j_1 k_1} \subset E_{j_2 k_2}$. Also, if $x \in E^a$, then $x \in E_{jk}$ for sufficiently large j and k. Thus, it follows that for such j and k, E_{jk} and E_a are not metrically separated. Then, by Exercise 6, there is a set $E \subset E_a$ with $|E| > 0$ at which the density of E_{jk} is greater than zero. Since f is measurable, f is approximately continuous at almost all points of A (Theorem 2.3.13). Therefore, there are points of E at which f is approximately continuous. Let $x \in E$ be a point of approximate continuity of f. Let c be a real number such that $c_j < c < a$. Let h' be a value of h such that $0 < h' < 1/2k$ and

$$(10.13) \qquad \frac{f(x + h') - f(x)}{h'} > c,$$

$x + h' \in A$. We can find such a number h' because $x \in E^a$. Let $\xi \in E_{jk}$ and take h satisfying (10.12) with $\xi + h = x + h'$. Then from (10.12) and (10.13), we get

$$f(x + h') - f(x) > h'c,$$

$$f(\xi + h) - f(\xi) < hc_j,$$

$\xi + h = x + h'$, $c_j < c$. These give

$$f(\xi) - f(x) > h'\left(c - \frac{h}{h'}c_j\right).$$

As $\xi \to x$, $h/h' \to 1$. Thus,

$$\liminf_{\xi \to x} [f(\xi) - f(x)] \geq h'(c - c_j),$$

and the set $\xi \in E_{jk}$ has density greater than zero at x. This contradicts the assumption that f is approximately continuous at x. Thus, our assumption that E^a is not measurable for some a is false; therefore, $_A D^+ f(x)$ is measurable on A.   □

Theorem 10.14. *If f is defined and finite on the set A, then the set of points $x \in A$ at which $_A D_- f(x) > {_A} D^+ f(x)$ is at most denumerable.*

Proof. Let $h < k$ be rational numbers and n be a positive integer. Let E_{hkn} be the set of points x such that

$$\frac{f(\xi) - f(x)}{\xi - x} < h,$$

$$\frac{f(\xi') - f(x)}{\xi' - x} > k$$

with $\xi - x < 1/n$ and $x - \xi' < 1/n$.

If $x_1 \in (x - 1/n, x + 1/n)$, $x_1 < x$, and $x_1 \in E_{hkn}$, then we can apply both of the above inequalities (the first with $\xi = x$, $x = x_1$ and the second with $\xi' = x_1$, $x = x$) and get

$$\frac{f(x) - f(x_1)}{x - x_1} < h,$$

$$\frac{f(x_1) - f(x)}{x_1 - x} > k.$$

Since $h < k$, this leads to a contradiction.

Thus, there is no $x_1 < x$, $x_1 \in (x - 1/n, x + 1/n)$ which is in E_{hkn} if $x \in E_{hkn}$. Hence, all $x \in E_{hkn}$ are isolated points and thus denumerable.

Next if x is a point such that $_A D_- f(x) > {_A} D^+ f(x)$, i.e.,

$$\liminf_{k \to 0-} \frac{f(x) - f(x - k)}{k} > \limsup_{h \to 0+} \frac{f(x + h) - f(x)}{h},$$

then x is in some E_{hkn}. But the set of triples (h, k, n) is denumerable. Hence, the set of points x at which $_A D_- f(x) > {_A} D^+ f(x)$ is at most denumerable.   □

Theorem 10.15. *If f is defined and finite on a set A, then the points $x \in A$ at which $_A D_+ f(x) > {_A} D^- f(x)$ is at most denumerable.*

Proof. The proof is similar to that of the preceding theorem.   □

Theorem 10.16. *If f is defined and finite on a set A, the set of points at which the right and left derivatives exist and are different is denumerable.*

Proof. Let

$$E = \{x: f'_-(x) < f'_+(x)\}.$$

Then for each $x \in E$, there exists a smallest rational number r_k such that

$$f'_-(x) < r_k < f'_+(x)$$

and a smallest integer m such that

(10.17)
$$\frac{f(\xi) - f(x)}{\xi - x} < r_k, \quad r_m < \xi < x, \, \xi \in A,$$

and a smallest integer n such that

(10.18)
$$\frac{f(\xi) - f(x)}{\xi - x} > r_k, \quad x < \xi < r_n, \, \xi \in A.$$

The triad associated with each x is unique because, if x_1 and x_2 $(x_1 < x_2)$ have the same triad, setting $\xi = x_1$ and $x = x_2$ in (10.17), we get

$$f(x_1) - f(x_2) < r_k(x_1 - x_2).$$

Setting $\xi = x_2$ and $x = x_1$ in (10.18) yields

$$f(x_2) - f(x_1) > r_k(x_2 - x_1).$$

This contradiction proves the uniqueness. Since such triads are denumerable, E is denumerable. That the set

$$E' = \{x: f'_-(x) > f'_+(x)\}$$

is denumerable is similarly proved. The desired result then follows. $\square$

Remark. Most of the results proved in Chapters 3 and 4 for the Dini derivatives and the approximate derivates of measurable functions defined on finite intervals have analogous results for *arbitrary functions defined on arbitrary sets*. The interested reader may consult [J1, pp. 181–203]. Most notably, the Denjoy–Saks–Young Theorem, characterizing the behavior of the Dini derivatives of a finite function defined on an interval, also holds for a function defined on an arbitrary set.

EXERCISES

1. Suppose

$$A = \bigcup I_i,$$
$$A = \bigcup I'_i$$

are two representations of the set A. Show that $|A|$ is independent of the representation of A.

2. For any two sets A and B, show

$$|A \cup B| \leq |A| + |B|.$$

3. Show that equality holds in Exercise 2 if A and B are metrically separated.

4. Prove Theorem 10.6.

5. Using Theorem 10.7 and the definition of measurability, show that if A is measurable, then at almost all points of A the density of A^C is zero and at almost all points of A^C the density of A is zero.

6. If A and B are not metrically separated, then $|A_B| = |B_A| > 0$, where A_B and B_A are the parts of A and B, respectively, at which the density of B and A is greater than zero.

References

[A1] Adams, C.R., The space of functions of bounded variation and certain general spaces, *Transactions of the American Mathematical Society*, 40 (1936), 421–438.

[AC] Adams, C.R. and Clarkson, J.A., On convergence in variation, *Bulletin of the American Mathematical Society*, 40 (1934), 413–417.

[AL] Adams, C.R., and Lewy, Hans, On convergence in length, *Duke Mathematical Journal*, 1 (1935), 19–26.

[A2] Apostol, Tom M., *Mathematical Analysis*, Addison-Wesley, Reading, MA, 1974.

[AB] Asplund, Edgar, and Bungart, Lutz, *A First Course in Integration*, Holt, Rinehart, and Winston, New York, 1966.

[A3] Austin, Donald, A geometric proof of the Lebesgue differentiation theorem, *Proceedings of the American Mathematical Society*, 16 (1965), 220–221.

[B1] Banach, Stefan, Sur les fonctions dérivées des fonctions mesurables, *Fundamenta Mathematicae*, 3 (1922), 128–132.

[B2] Bruckner, Andrew M., *Differentiation of Real Functions*, Springer-Verlag, Berlin, 1978.

[BH1] Buchanan, H.E. and Hildebrandt, T.H., Note on the convergence of a sequence of functions of a certain type, *Annals of Mathematics, Series 2*, 9 (1908), 123–126.

[BH2] Burkhill, J.C. and Haslam-Jones, U.S., The derivatives and approximate derivatives of measurable functions, *Proceedings of the London Mathematical Society (2)*, 32 (1931), 346–355.

[C1] Clarkson, J.A., A property of derivatives, *Bulletin of the American Mathematical Society*, 53 (1947), 124–125.

253

[D1] deBarra, G., *Measure Theory and Integration*, John Wiley and Sons, New York, 1981.

[D2] Dunford, Nelson, On a theorem of Plessner, *Bulletin of the American Mathematical Society*, 41 (1935), 356–358.

[F1] Feller, William, *An Introduction to Probability Theory and Its Applications*, Vol. 1, John Wiley & Sons, New York, 1957.

[G1] Gilman, R.E., A class of functions continuous but not absolutely continuous, *Annals of Mathematics, Series 2*, 33 (1932), 433–442.

[G2] Goffman, Casper, "A geometric characterization of bounded variation and absolute continuity," *Bulletin of the Institute of Mathematics, Academia Sinica*, 9(3) (1981), 395–398.

[G3] Goffman, Casper, A remark on absolutely continuous functions, *Real Analysis Exchange*, 15 (1989–1990), 384–385.

[G4] Goffman, Casper, Lusin type theorems for functions of bounded variation, *Real Analysis Exchange*, 5 (1979–1980), 261–266.

[G5] Goffman, Casper, On functions with summable derivatives, *American Mathematical Monthly*, 78 (1971), 874–875.

[G6] Goffman, Casper, *Real Functions*, Rinehart & Company, New York, 1953.

[GN] Goffman, Casper and Neugebauer, C.J., On approximate derivatives, *Proceedings of the American Mathematical Society*, 11 (1960), 962–966.

[GP] Goffman, Casper and Pedrick, George, *First Course in Functional Analysis*, Prentice-Hall, Englewood Cliffs, N.J., 1965.

[GW] Goffman, Casper and Waterman, Daniel, Approximately continuous transformations, *Proceedings of the American Mathematical Society*, 12 (1961), 116–121.

[HB] Herzog, F. and Bissinger, B.H., A Cantor function constructed by continued fractions, *Bulletin of the American Mathematical Society*, 53 (1947), 104–115.

[H1] Hildebrandt, T.H., Necessary and sufficient conditions for the interchange of limit and summation in the case of sequences of infinite series of a certain type, *Annals of Mathematics, Series 2*, 14 (1912–1913), 81–83.

[HT] Hille, E. and Tamarkin, J.D., Remarks on a known example of a monotone continuous function, *American Mathematical Monthly*, 36 (1929), 255–264.

[H2] Hobson, E.W., *The Theory of Functions of a Real Variable and the Theory of Fourier Series*, Vol. 1, 3rd ed., reprinted by Dover Publications, New York, 1957.

[H3] Hsiang, Fu Cheng, On differentiable functions, *Bulletin of the American Mathematical Society*, 66 (1960), 382–383.

[J1] Jeffery, R.L., *The Theory of Functions of a Real Variable*, 2nd ed., University of Toronto Press, Toronto, 1953.

[J2] Jones, Frank, *Lebesgue Integration on Euclidean Space*, Jones and Bartlett Publishers, Boston, 1993.

[K1] Kestelman, H., An integral for functions of bounded variation, *Journal of the London Mathematical Society*, 9, (part 3) (1934), pp. 174–178.

[K2] Kestelman, H., *Modern Theories of Integration*, Dover Publications, New York, 1960.

[K3] Kober, H., On singular functions of bounded variation, *Journal of the London Mathematical Society*, 23 (1948), 222–229.

[M1] Marcus, Solomon, On a theorem of Denjoy and on approximate derivative, *Monatshefle fuer Mathematik*, 66 (1962), 435–440.

[M2] McShane, Edward James, *Integration*, Princeton University Press, Princeton, 1944.

[M3] Morse, Anthony, Convergence in variation and related topics, *Transaction of the American Mathematical Society*, 41 (1937), 48–83.

[M4] Munroe, M.E., *Introduction to Measure and Integration*, Addison-Wesley, Reading, MA, 1953.

[N1] Natanson, I.P., *Theory of Functions of a Real Variable*, Vol. 1, rev. ed., Frederick Ungar Publishing, New York, 1961.

[PCP] Pu, H.H., Chen, J.D., and Pu, H.W., A theorem on approximate derivates, *Bulletin of the Institute of Mathematics Academia Sinica*, 2(1) (1974), 87–91.

[PP1] Pu, H.W. and Pu, H.H. On a generalized Denjoy property of approximate derivatives, *Bulletin de la Société Royale des Sciences de Liège*, 46(9–10) (1977), 222–223.

[PP2] Pu, H.W., and Pu, H.H., On a certain property of approximate derivative, *Bulletin de la Société Royale des Sciences de Liège*, 46(5–8) (1977), 162–166.

[R1] Randolph, John F., *Basic Real and Abstract Analysis*, Academic Press, New York, 1968.

[R2] Randolph, John F., Distances between points of the Cantor set, *American Mathematical Society Monthly*, 47 (1940), 549–551.

[R3] Riesz, F. and Sz.-Nagy, Béle, *Functional Analysis*, Dover Publications, New York, 1990.

[R4] Royden, H.L., *Real Analysis*, 3rd ed., Macmillan Publishing, New York, 1988.

[R5] Rudin, Walter, *Real and Complex Analysis*, McGraw-Hill, New York, 1966.

[R6] Ruziewicz, Stanislaw, Sur les fonctions qui ont la même dérivée et dont la différence n'est pas constante, *Fundamenta Mathematicae*, 1 (1920), 148–151.

[S1] Saks, Stanislaw, *Theory of the Integral*, 2nd rev. ed., Hafner Publishing, New York, 1937.

[S2] Salem, R., On some singular monotonic functions which are strictly increasing, *Transactions of the American Mathematical Society*, 53 (1943), 427–439.

[S3] Sierpiński, W., Sur une généralisation de la notion de la continuité approximative, *Fundamenta Mathematicae*, 4 (1923), 124–127.

[T1] Takacs, Lajos, An increasing continuous singular function, *American Mathematical Monthly*, 85 (1978), 35–37.

[T2] Taylor, Angus E., *General Theory of Functions and Integration*, Dover Publications, New York, 1965.

[T3] Taylor, Angus E., *Introduction to Functional Analysis*, John Wiley & Sons, New York, 1958.

[T4] Temple, G., *The Structure of Lebesgue Integration*, Oxford at the Clarendon Press, Oxford, 1971.

[T5] Tolstoff, G., Sur la dérivée approximative exact, *Recueil Mathématique (Matematicheskici Sbornik)*, 4 (1938), 499–504.

[T6] Tolstoff, G., Sur quelques propertiés des fonctions approximativement continues, *Recueil Mathématique (Matematicheskici Sbornik)*, 5 (1939), 632–645.

Additional References

The authors are indebted to the following sources for many of the exercises in this book.

Boas, Ralph P., Jr., *A Primer of Real Functions*, Quinn and Boden Co., Rahway, N.J., 1972.

George, Claude, *Exercises in Integration*, Springer-Verlag, New York, 1984.

Hewitt, Edwin and Stromberg, Karl, *Real and Abstract Analysis*, Springer-Verlag, New York, 1965.

Shilov, G. Ye., *Mathematical Analysis*, Pergamon Press, Oxford, 1965.

Stromberg, Karl, *An Introduction to Classical Real Analysis*, Wadsworth, Inc., Belmont, Ca., 1981.

Varberg, Dale, On absolutely continuous functions, *American Mathematical Monthly*, 72 (1965), 831–841.

Index

Universitext   *(continued)*